Real-Time Management of Resource Allocation Systems

REAL-TIME MANAGEMENT OF RESOURCE ALLOCATIONS SYSTEMS
A Discrete Event Systems Approach

SPYROS A. REVELIOTIS
School of Industrial & Systems Engineering
Georgia Institute of Technology
Atlanta, GA 30032

 Springer

Spyros A. Reveliotis
Georgia Institute of Technology
Atlanta, GA, USA

Library of Congress Cataloging-in-Publication Data
A C.I.P. Catalogue record for this book is available
from the Library of Congress.

ISBN 0-387-23960-X   e-ISBN 0-387-23967-7   Printed on acid-free paper.

Printed in the United States of America.

9 8 7 6 5 4 3 2 1   SPIN 11054313

springeronline.com

Contents

Preface

This book deals with the problem of managing the resource allocation that takes place in the operational context of various contemporary technological applications, including flexibly automated production systems, automated railway and/or monorail transportation systems, electronic workflow management systems and business transaction supporting systems. A distinguishing trait of all the aforementioned applications, is that they seek to limit the role of the human element to remote high-level supervision, while placing the burden of the real-time monitoring and coordination of the ongoing activity upon a computerized control system. This development is justified by a number of technical, economic and safety considerations, and it is facilitated by the advent of modern computing and sensing technologies. On the other hand, a challenging task for the effective deployment of these target applications, is the synthesis of the *control logic* that will manage the allocation of the limited system resources to the concurrently running processes; this logic must guarantee the correct and expedient execution of all the active processes, while minimizing the need for external human intervention. The resulting problem is rather novel for the developers of these systems, since, in the past, many of its facets were left to the jurisdiction of the present human intelligence. It is also complex, due to the high levels of choice – otherwise known as *flexibility* – inherent in the operation of these environments. As a result, most of the resource allocation controllers currently developed for the automated versions of the aforementioned applications, are based on ad-hoc and simplistic solutions; unable to deal systematically with the behavioral complexity of the underlying system, these solutions, in their effort to ensure correct and robust operation, constrain unnecessarily the inherent operational flexibility, and eventually, they compromise the overall efficiency and productivity.

Yet, the last 10-15 years have seen the development of a series of results, coming primarily from the burgeoning area of *Discrete Event Systems (DES)*, that have enhanced our understanding of the behavioral properties of the aforementioned applications with respect to the underlying resource allocation function, and they offer a solid theoretical basis towards the formal representation, analysis and control of this function. *The main objective of this book is the systematic exposition of these results, and their orchestration in a complete control framework, appropriate for the aforementioned environments.* Hence, the book opens with an introductory description of (some of) the target application environments, and the underlying resource allocation function. This description leads to the specification of the control requirements, and the outline of a control framework that can provide a viable solution to these requirements. A notable feature of the proposed control framework is the emphasis that it places on (i) the *robustness* of the deployed control function with respect to the system stochasticities and the various operational contingencies, (ii) the *scalability* of the proposed solutions, so that they apply to the large-scale context of the target technological applications, and (iii) the *operational efficiency* of the resulting controlled system. These three properties are supported through the adoption of a *"closed-loop"* structure for the proposed control scheme, and also, through a pertinent *decomposition* of the overall control function, to one component seeking the logical correctness and consistency of the system behavior, and another component addressing performance considerations; the first of these components is characterized as the *logical controller* of the underlying resource allocation function, while the second component is characterized as the *performance-oriented controller*. The rest of the book is devoted to a rigorous study of the control problems addressed by each of these two controllers, and their integration to a unified control function. A notion of *optimal control* is formulated for each of these problems, but it turns out that the corresponding optimal policies are computationally intractable. Hence, a large part of the book is devoted to the development of effective and computationally efficient *approximations* for these optimal control policies, especially for that corresponding to the more novel logical control problem.

The above paragraphs have outlined the "utilitarian" aspects of the book content. However, the book is intended to play a number of additional roles, each defined with respect to a different audience: Hence, for the Industrial Engineering / Operations Research-oriented student, researcher and sophisticated practitioner, the book offers an introduction to the broader area of Discrete Event Systems, and familiarization with an array of its main modelling frameworks, concepts and algorithms. Furthermore, this material is provided through implementation on a practical application context, that reveals the potential of this fairly novel area, and expands our analysis and design capability for a class of problems that traditionally has belonged to the domain of the IE/OR disciplines. For the student and researcher of the area of Discrete Event Systems itself, the book offers a contextual, in-depth application of many concepts and algorithms developed by DES theory, but even more importantly, it stresses the complications arising from the non-polynomial complexity that is inherent in many of these concepts and algorithms, and addresses the issue of developing effective and computationally efficient / polynomial approximations. Indeed, a major contribution of this book is that it epitomizes a significant body of results, that were developed over the last 12 years, and concern the design of provably correct and computationally efficient solutions for the considered logical control problem. Finally, a last intention of the book is to highlight the missing links in the presented developments, and to motivate, in this way, further research activity by the interested communities.

From a presentational standpoint, there has been an effort to keep the book development as independent as possible, but the reader is expected to be familiar with some fundamental mathematical concepts, like the concepts of a graph or a probability distribution. In each chapter, an opening section summarizes all the background material necessary for the subsequent developments, and it provides a number of citations for readers that would like to have a more in-depth or leisurely treatment of this material. On the other hand, the development of the main results of each chapter has placed the emphasis on the rigor, conciseness, and also, the lucidity of the exposition. The selection of the material included in each chapter has been aligned with and supports the posed specifications of robustness, scalability and efficiency for the proposed solution; however, a closing section at each chapter overviews the broader set of available results on the problems considered in it, and it provides a "historical perspective" for the presented developments. The book can function as the basis for a senior-level undergraduate or graduate-level course on resource allocation systems and their real-time management, or some more "concrete" version of this problem defined in the context of the aforementioned target application environments. It would also constitute a good supplement for any introductory course on Discrete Event Systems theory.

Concluding this expository discussion of the book and its content, I would like to take the opportunity to acknowledge and thank a number of parties that have been valuable in the development of this work. Hence, first I would like to thank Placid Ferreira, Mark Lawley, Jonghun Park and Jin Young Choi, whose collaboration and friendship have been very instrumental in the development of many of the results presented in this book. I would also like to extend many thanks to all my colleagues who, through personal communication, conference participation, or simply through their own personal contributions, have been great inspiration and partners in the entire journey that has culminated to the writing of this book. I am particularly indebted to Elzbieta Roszkowska who read many parts of the manuscript and provided valuable feedback during the book development. The School of Industrial Engineering at Georgia Tech, and especially, the Virtual Factory Lab in it, have been a very supportive environment for the research activity that has led to many of the results reported in this book.

Furthermore, the Virtual Factory Lab and the Keck Foundation, together with NSF and The Logistics Institute - Asia Pacific, have provided the funding for a substantial part of this research; I am grateful to them. Finally, I want to thank Fred Hillier, Gary Folven and Kluwer for the encouragement they provided towards undertaking the writing of this book, and for "hosting" the book in their prestigious series.

SPYROS A. REVELIOTIS

Chapter 1

RESOURCE ALLOCATION SYSTEMS: CONCEPTS AND PROBLEMS

As indicated by its title, this book is about *resource allocation*. Admittedly, this is a very general term, already extensively used in the scientific literature, in a number of application contexts, and with various technical meanings and connotations. Hence, in this introductory chapter, we shall try, through a number of examples, to specify the resource allocation systems (RAS) considered in this work, and to motivate the issues and problems that will be the subject of our study. We shall also outline the basic methodological framework that provides the context for the more detailed analysis undertaken in the subsequent chapters. A third objective of this introductory discussion is to emphasize the particular elements that differentiate the methodological approach adopted in this work from other approaches already advocated in the literature as potential solutions to the problems under consideration. The chapter will conclude with a brief discussion of the book organization and its major contributions.

1.   Some example resource allocation systems

The following examples provide a tangible characterization of the operational systems considered in this work and the underlying resource allocation. Coming from a broad range of application domains, these examples reveal the ubiquitous and permeating nature of the undertaken problems, and, hopefully, provide the motivation for their study and evidence for the potential value of the derived results.

Example 1: A flexibly automated robotic cell.   Historically, the ability to support the effective and efficient operation of flexibly automated production systems, has been one of the key motivations and drivers of the work presented in this book. Generally speaking, a flexibly automated production system consists of a number of manufacturing workstations of finite buffering and processing ca-

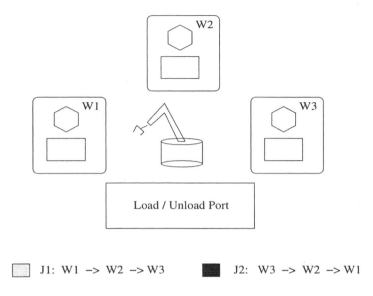

J1: W1 –> W2 –> W3   J2: W3 –> W2 –> W1

Figure 1.1. A flexibly automated robotic cell

pacity, interconnected by an automated material handling system. Furthermore, the material handling system is of *unit-load* type, i.e., it consists of a number of transporters, with each transporter being able to carry one work-piece at a time. In particular, for fairly small and highly concentrated configurations, the material handling function can be supported by a number of robotic manipulators, while for more spatially dispersed systems, this function is supported by an *Automated Guided Vehicle (AGV)* system or an *overhead monorail* system.[1] In its typical operation, such an integrated production environment supports the production of a number of different parts, that are manufactured through a sequence of processing stages. Each stage is carried out at some of the system workstations while parts are transferred from station to station automatically, through the aforementioned material handling system. The notion of flexibility in such an environment pertains to its ability to support the simultaneous production of more than one part, each of them possibly executing a distinct processing sequence.

As a case in point, consider the small robotic cell of Figure 1.1. This cell consists of three workstations, W_1, W_2 and W_3, each being able to hold one part at a time, and a loading port, LP, where parts are introduced into and removed from the system. The part transfer among the various system stations and / or

---

[1]A particular variation of these systems is discussed more extensively in the next example.

the loading port is facilitated by a robotic manipulator, able to carry one part at a time. In the considered configuration, the cell supports the production of two part types, J_1 and J_2. Each part type visits all three workstations, however, the corresponding sequences differ; specifically, the processing sequence for part type J_1 is $W_1 \rightarrow W_2 \rightarrow W_3$, while the processing sequence for part type J_2 is $W_3 \rightarrow W_2 \rightarrow W_1$.

In order to characterize the operation of the aforementioned environment as a resource allocation system of the type considered in this work, we need to identify two primary constituent elements:

  i the set of *resource types* and their corresponding *capacities*, i.e., the number of distinct identical units available from each resource type;

  ii the set of *process types*, defined by the *sequences* of their *processing stages*, and the *resource allocations* and *timings* associated with these stages.

Strictly speaking, in the context of the considered example, the distinct resource types are (i) the robotic manipulator, (ii) the workstation buffers, and (iii) the workstation processors, each available at a single unit of capacity. However, the single unit of buffering capacity at each workstation implies that a job being allocated this capacity will also be naturally allocated the workstation processor; hence, it is pertinent to "bundle" the processor and the buffer at each workstation W_i, $i = 1, 2, 3$, to a single resource, to be denoted by R_i. Furthermore, we shall use R_0 to denote the resource corresponding to the robotic manipulator. All resources R_i, $i = 0, \ldots, 3$, are *reusable*, i.e., they can be acquired, utilized and released by the various processes, without each of these allocation cycles affecting their future availability. With this definition of the system resources in place, the system process types, corresponding to the two distinct parts, J_1 and J_2, can be respectively characterized by the following resource allocation sequences:

- $\Pi_1 : R_0 \rightarrow R_1 \rightarrow R_0 \rightarrow R_2 \rightarrow R_0 \rightarrow R_3 \rightarrow R_0$

- $\Pi_2 : R_0 \rightarrow R_3 \rightarrow R_0 \rightarrow R_2 \rightarrow R_0 \rightarrow R_1 \rightarrow R_0$

Each process Π_j, $j = 1, 2$, consists of a sequence of seven stages, Ξ_{jk}, $k = 1, \ldots, 7$, interleaving transport and processing operations, and with each stage Ξ_{jk} requesting the single unit of some resource R_i, $i = 0, \ldots, 3$, for its execution. This latter effect could have been equivalently expressed by associating with each processing stage Ξ_{jk} a *resource allocation vector* $A_{jk} = \mathbf{1}_{r(j,k)}$, where $r(j, k) \in \{0, \ldots, 3\}$ denotes the resource type engaged by stage Ξ_{jk}, and $\mathbf{1}_l$ denotes, in this context,[2] the 4-dimensional unit vector with the unit ele-

---

[2]More generally, in this book, $\mathbf{1}_l$ will denote the unit vector with the unit element in its l-th component, while the vector dimensionality will be determined by the discussion context. Similarly, $\mathbf{1}$ will denote the vector with all components equal to 1, and $\mathbf{0}$ will denote the vector with all components equal to 0.

ment at the l-th component. This vector-based representation of the underlying resource allocation, while more cumbersome in the context of the considered example, provides the definitiveness and generality to express the resource allocation underlying more general situations, where a process stage might request more than one resource type and/or more resource units from each resource type. The resource allocation model characterizing the operation of the considered robotic cell is completed, for the purposes of our analysis, by providing for each processing stage Ξ_{jk}, a timing distribution D_{jk} that characterizes the transfer and/or processing times t_{jk} experienced during the execution of that stage.

The basic problem that will be addressed by this work in the context of the robotic cell described above, is the management of (i) the loading of the various job instances into the cell, and (ii) their advancement through their corresponding processing sequences, so that a number of operational objectives are satisfied. The Operations Research (OR) oriented reader will immediately recognize that some such objectives typically identified and studied in the context of the more "traditional" OR-related frameworks, are defined by the optimization of some *performance index* like the maximization of the (long run) system *throughput* – i.e., the average number of parts produced per unit of time, the minimization of the expected part *cycle times* – i.e., the average time it takes for a part to run through the cell, and/or the minimization of some average *work-in-process (WIP) inventory cost* function. The resulting problem formulations belong to the class of *scheduling* and *job sequencing* problems that constitutes part of the broader OR literature. The work presented in this book addresses these performance-related issues, too, but it *complements* them with an additional class of problems that aims at *generating "correct" system behavior*; this issue is especially important given the automated mode of operation of the targeted application environments. To exemplify this additional class of problems, consider the robotic cell under the resource allocation state presented in Figure 1.2. In the depicted state, workstation W_1 is occupied by a job instance executing stage Ξ_{11} and workstation W_2 is occupied by a job instance executing stage Ξ_{22}. It should be obvious to the reader that no matter which of these two jobs is picked by the robot for transferring to its next requested resource, the system will get stalled, since upon reaching the next workstation, the robot will not be able to deposit the job. Under automated operation, this problem will persist until it is recognized and resolved through some exception handling routine, most probably involving external human intervention. The problematic situation depicted in Figure 1.2 is known as a (manufacturing) system *deadlock*, and its effective resolution is going to be a major theme of this book.

To further motivate the environments and problems considered in this work, it is worth-noticing that the robotic cell structure outlined in this example is in

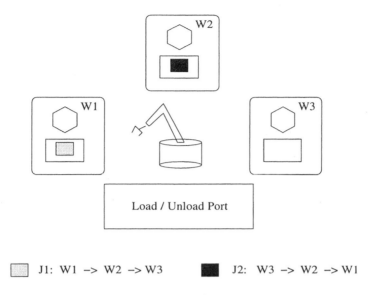

J1: W1 –> W2 –> W3   J2: W3 –> W2 –> W1

Figure 1.2. A manufacturing system deadlock

direct correspondence to the *cluster tool* (Singer, 1995) topology that currently is used extensively in the semiconductor manufacturing industry. In the cluster-tool micro-environment, the different workstations correspond to the *chambers* mounted on the tool, with each chamber supporting a distinct process, while the parts circulating through the tool are the single *wafers* contained in the *wafer cassette* staged at the tool load-port. Under the current industry practice, the aforementioned deadlock-related resource allocation problems either are addressed through some control scheme developed in a totally ad-hoc fashion, or, most typically, are eliminated by ensuring that each cassette wafer contains a single part type executing a single operation at each tool chamber. Although quite robust from an operational standpoint, this last practice enforces a *batch*-based operational mode, and it stifles the system flexibility, responsiveness and potential productivity. In other words, it can be argued that, in its effort to ensure correct behavior, the current industry practice underlying the operation of these environments, defeats some of the key objectives and operational aspects underlying the notion of flexible automation, and it fails to materialize the frequently advertised benefits of this technology.

Example 2: A zone-controlled AGV system. As previously mentioned, *Automated Guided Vehicle (AGV)* systems are considered as one of the most appropriate modes for material handling support of contemporary flexibly au-

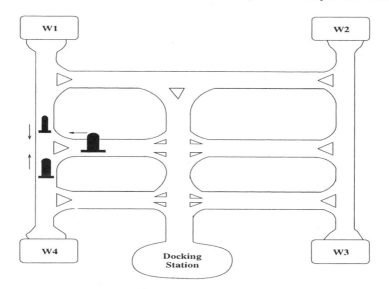

Figure 1.3. An AGV system deadlock

tomated production environments. According to (Ganesharajah et al., 1998), some of the primary advantages attributed to these environments are increased routing flexibility, space utilization and safety, and reduced overall operational cost.

The basic structure of the AGV system typically used in contemporary manufacturing environments is depicted in Figure 1.3. In general, such a system consists of a set of cooperating driver-less vehicles which transport goods and materials among the different workstations and storage sites of a production facility, by following a set of predetermined, physical or virtual guidepaths embedded in the facility layout, and coordinated by a centralized or distributed computer-based control system. Motion on the different links of the guidepath network is *bidirectional*. However, the AGV's themselves are assumed to be *unidirectional*, i.e., they must make a U-turn in order to travel in the opposite direction.[3] To avoid physical collision, *zone control* is applied; i.e., the guide-path network is partitioned into segments with each segment being accessed by only one vehicle at a time. A natural segmentation associates one zone with each intersection and workstation node, and each path link. In case, however, that there are lengthy path links, they can be further segmented into a series of

---

[3]In addition to expressing the current capabilities of AGV technology, this constraint also disables *backtracking*, and therefore, it enforces a sense of intention in the vehicle motion.

zones. Finally, the system possesses some *docking* station(s) where idle vehicles park and potentially recharge their batteries. The capacity of the docking station(s) is sufficient to accommodate all the vehicles, e.g., when the system is shut down.

A vehicle trip in the considered AGV system can be conceptualized as a "mission" that starts from the docking station, visits successively a *"source"* and a *"destination"* workstation, and eventually concludes again at the docking station. It is possible, however, – in fact, quite probable – that a vehicle having completed its trip up to the "destination" point, will be re-assigned and redirected to another trip, before reaching the docking station. Yet, the conceptual requirement that an *idle* vehicle should (be able to) eventually return to the docking station is important for fitting the operation of the aforementioned AGV system in the RAS framework considered in this work; it introduces an *episodic* decomposition of the overall vehicle movement, enabling, thus, the correspondence of each such episode to an active system *process*. In other words, active processes in the RAS modelling the considered AGV system are defined by the running vehicle missions. The *resources* utilized by each such process are the various guidepath zones occupied by the vehicle during its travelling among the various milestone nodes of its trip. Under *static* vehicle routing, the sequence of zones utilized by a vehicle when travelling between any two milestone nodes is determined a priori – e.g., based on some "shortest path" computation. When *dynamic* vehicle routing is applied, the exact vehicle route is determined on-line, in a way that takes into consideration the congestion status of the entire network.

From a control-theoretic standpoint, the key issues to be addressed in the real-time management of the aforementioned AGV system are:

i the matching of transfer requests with the system vehicles, i.e., the assignment of a pending transfer request to a vehicle completing its current mission, or the assignment of an idle vehicle to an emerging transfer request;

ii the efficient and conflict-free routing of the currently active vehicles towards their target nodes.

Under the RAS interpretation of the system dynamics adopted in this work, the first of the above issues corresponds to the decision of *loading* a new process into the system. The second issue corresponds to the *sequencing* decisions underlying the advancement of the various active processes towards their completion. Furthermore, similar to the robotic cell case, these decisions are driven by, both, performance and behavioral correctness concerns. Specifically, while the enforced zone control prevents the physical collision of the vehicles, careless routing can still give rise to situations similar to those depicted in Figure 1.3, where a set of vehicles persistently blocks each other's way; this type of deadlocking situation is characterized as an *AGV deadlock*.

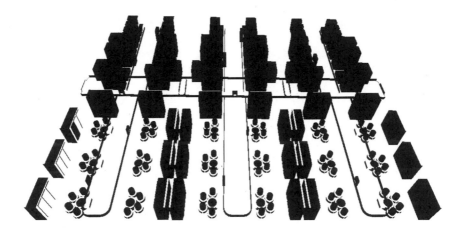

Figure 1.4.   The typical lay-out of a modern semiconductor fab

It is interesting to notice that the current industry practice with respect to the AGV deadlock is similar to that applied to the manufacturing system deadlock; i.e., the problem is eliminated at the system design level by constraining the vehicle motion through the adopted topology for the guidepath network. Specifically, the currently adopted *tandem* configurations divide the overall guidepath network to a number of non-overlapping unidirectional loops, interfaced with some strategically positioned exchange buffers; Figure 1.4 depicts the implementation of this idea in the context of the contemporary semiconductor fabs, (Rust et al., 2002). Such a solution prevents the development of AGV deadlocks, but it introduces a cumbersome and time consuming "hand-over" process at the various exchange buffers, and, in certain cases, it enforces unnecessarily long trips, since the vehicles have to always travel in the preselected unidirectional sense over the loop, even in the case that the destination station is located close to the source station, but in the opposite direction. Furthermore, the traffic in each loop tends to experience extensive congestion, since it is paced by the slowest ongoing activity (e.g., in the overhead monorail system of the fab depicted in Figure 1.4, a vehicle loading or unloading a part will block any other vehicle behind it in the loop). The conclusion remains the same: ensuring operational correctness in a simple-minded manner, will eventually stifle the system operational flexibility and lead to considerable inefficiency.

Figure 1.5. An urban monorail system

Example 3: An urban monorail system. The previous discussion regarding the control of AGV systems can be naturally extended to the control of urban monorail systems similar to the one depicted in Figure 1.5. Such systems are popular for interconnecting airport terminals, but they have also been deployed in certain city centers, e.g., downtown Seattle and Sydney. Similar to the AGV case, currently, the traffic flows supported by these systems are extremely simple: either a unidirectional loop or a single vehicle moving back and forth on some line segment. The theory developed in this book would facilitate conflict-free and efficient traffic over arbitrary guidepath networks. In a more futuristic setting, one can also envision a complex elevator system where the elevator cells are moving through a three-dimensional grid in order to transfer passengers among a number of access locations dispersed over an entire city block.[4]

Example 4: A computerized workflow management system. Computing technology has evolved to the point that it permeates every transactional and administrative activity in contemporary organizations, from business, to government, to military operations. One of the key features that underlies the wide

---

[4] Such a prototypical project concerning the design, manufacture and operation of tomorrow's hyper-buildings is currently under development in Japan; c.f., *Tokyo's Sky City*, Discovery Channel web-page.

acceptance of this technology in all the aforementioned applications, and has driven to a large extent its past growth, is the effective separation of the *generic* application logic from the *case-specific* computational data. This separation provides extensive versatility and re-usability to contemporary applications of information technology (IT), and has driven the inception, development and growth of very successful concepts and industries, like the database-related software and applications, and object-oriented technologies.

Presently, the advent of distributed and network – including the Internet – based computing has enabled the integration and co-ordination of organizational activities that are *functionally* and/or *geographically* distributed across the entire domain of the considered operational environment. However, the deployment of such distributed organizational activity networks is currently focused only on the technologies necessary for Internet/Intranet-based data exchange and message passing; the logic applied for the activity-flow control and co-ordination is of very ad-hoc nature, frequently introduced only in the corporate written policies and procedures. It is becoming, thus, increasingly clear that, in order to fully exploit the capabilities and advantages of distributed and network-based computing, it is necessary to develop another kind of separation principle, this time isolating the computational component that defines and manages the organizational processes and the resulting workflow, from the application software supporting the various process stages, and the interfacing technology enabling the communication and interaction among the computational applications and/or the human element supporting the execution of the various processing stages. The resulting paradigm and technology is evolving around the concept of *Workflow Management System (WFMS)* (Lawrence, 1997), i.e., a computing system that completely defines, creates and manages the execution of workflows through the use of software running on one or more *workflow engines*; the latter are able to interpret the process definition, interact with the workflow participants, and where required, invoke and administer the use of IT tools and applications.

The basic structure and functionality of a typical WFMS is depicted in Figure 1.6. As indicated in Figure 1.6, a primary concept in the study and characterization of workflows is the workflow *process*, which is defined as a coordinated (parallel and/or serial) set of atomic process activities, that are connected in order to achieve a common business goal. The *process activity*, itself, is defined as a logical step or description of a piece of work that contributes toward the accomplishment of a process, and its execution requires the explicit allocation of a well-defined set of the system *resources*. These resources can range from IT tools and application software, to data-files, to the organization personnel possessing a particular set of skills – known as *organizational roles*, in the relevant parlance. A workflow process is first specified using a *process*

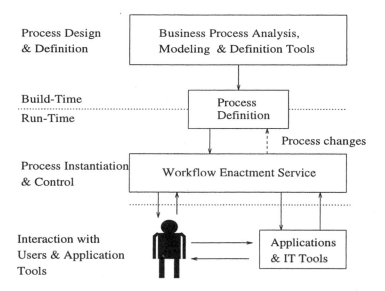

Figure 1.6. The basic structure and functionality of a Workflow Management System

definition formalism/language, and it is subsequently executed by a *workflow management system (WFMS)*. In the context of the broader discussion pursued in this chapter, the functionality of the typical WFMS software can be further distinguished to:[5]

Build-time functions that are concerned with the definition of the workflow processes and their constituent activities, and the verification of logical correctness and consistency of the underlying process logic;

Run-time control functions that are concerned with the orderly execution of the various process instances – workflow *cases*, in the corresponding terminology – according to the corresponding process-defining logic, and the establishment of deadlock-free and expedient resource allocation among the contesting cases.

It should be obvious to the attentive reader that the WFMS specification provided above essentially corresponds to a resource allocation function similar to

---

[5]Some additional functions supported by a typical WFMS are the facilitation of (i) the run-time interaction between the WFMS itself and the human users and IT application tools involved in the execution of the various process steps, and of (ii) the interfacing and information exchange between the various parties involved. However, these functions are beyond the scope of the control-theoretic perspective taken in this work, and they are addressed by the relevant computing and communication technologies.

that identified in the operation of the robotic cell and the AGV system discussed in the previous examples, and with the same set of concerns driving the underlying decision making process. We conclude the discussion of this example by noticing that, as an industry, *workflow management (WFM)* is a fast growing technology, that is increasingly applied in applications like insurance, banking, legal and general administration services. Furthermore, a considerable number of vendors offer a series of WFM products, and there is a continual introduction of more products into the market. We reiterate, however, that the current industry developments, when considered from a control-theoretic standpoint, are of a rather ad-hoc nature, lacking a well-defined modelling and analysis framework, that would systematically address the underlying design and operational issues. We expect and hope that the material developed in this book will contribute towards the development of the formal framework necessary to address these design and control requirements.

Example 5: Internet-based computing.   We conclude the motivational discussion for the resource allocation systems and problems considered in this book with a rather futuristic example taken from the world of the internet and its emerging functions. A concept currently quite popular in this domain is that of the *information grid*, i.e., the seamless integration of a large collection of computers to a virtual platform immediately accessible through the internet and providing a broad series of computational and information / data services, that could not have been supported independently by any of the grid nodes. The successful implementation of such a concept necessitates (i) the decomposition of any submitted job to a pertinent set of processing steps, (ii) the sequencing and coordination of these steps for their orderly execution, (iii) the matching of these steps with the available grid nodes / resources, on the basis of their processing requirements and the nodal capabilities, and (iv) the final accumulation and synthesis of the derived results in a user-friendly file document that will be presented to the grid user. Given the novelty of the application, and for the sake of brevity, we shall not expand any further in the description of this particular application, but we refer the interested reader to ((Marinescu, 2002), Section 5.8) for a more extensive introductory discussion. We notice, however, that the eventual implementation of this idea would give rise to many of the sequential resource allocation elements encountered in the application contexts considered in the previous examples.

2.   Sequential RAS and their Supervisory Control problem

Recapitulating the most salient aspects in the operation of the systems considered in the previous examples, we notice that all of them can be abstracted as a finite set of reusable resources that are exclusively allocated to a number of concurrently executing processes for the sequential execution of their various processing stages. Furthermore, the finiteness of the system resources when combined with the flexible mode of operation of the considered systems, necessitates the arbitration of the underlying resource allocation function in order to ensure (i) its logical correctness and inherent consistency, and (ii) its efficiency with respect to some stated performance objective(s). In this section, first we provide a formal characterization of the (sequential) resource allocation system (RAS), that will be the fundamental modelling abstraction employed in this work, and subsequently we discuss the primary control problems that must be addressed for the effective and efficient operation of any given RAS. This discussion will reveal the issues and concerns that must be addressed by any pertinent solution approach to the problem of real-time management of sequential resource allocation systems, and it will set the stage for the next section, which outlines the basic methodological framework adopted by the work presented in this book.

2.1 Resource Allocation Systems: Concepts and Definitions

The following definition of sequential resource allocation systems formalizes all the key concepts and insights provided by the example applications of the previous section, and it establishes the unifying framework that will be employed in the subsequent analysis:

DEFINITION 1 *A sequential resource allocation system (RAS) is defined as an 5-tuple* $\Phi = <\mathcal{R}, C, \mathcal{P}, \mathcal{A}, \mathcal{T}>$ *where:*

1 $\mathcal{R} = \{R_1, \ldots, R_m\}$ *is the set of the system* resource types.

2 $C : \mathcal{R} \to Z^+$ *– the set of strictly positive integers[6] – is the system* capacity *function, characterizing the number of identical units from each resource type available in the system. Resources are considered to be* reusable, *i.e., each allocation cycle does not affect their functional status or subsequent availability, and therefore,* $C(R_i) \equiv C_i$ *constitutes a system* invariant *for each i.*

3 $\mathcal{P} = \{\Pi_1, \ldots, \Pi_n\}$ *denotes the set of the system* process types *supported by the considered system configuration. Each process type* Π_j *is a composite element itself, in particular,* $\Pi_j = <\mathcal{S}_j, \mathcal{G}_j>$, *where:*

(a) $\mathcal{S}_j = \{\Xi_{j1}, \ldots, \Xi_{j,l(j)}\}$ *denotes the set of* processing stages *involved in the definition of process type* Π_j, *and*

(b) $\mathcal{G}_j$ *represents some data structure communicating some sequential logic that applies to the execution of any process instance of type* Π_j.

---

[6] Also, in this book, Z_0^+ will denote the set of nonnegative integers, Z will denote the set of all integers, and $\Re$ will denote the set of reals.

4 $\mathcal{A}$: $\bigcup_{j=1}^{n} \mathcal{S}_j \rightarrow \prod_{i=1}^{m}\{0,\ldots,C_i\}$ *is the* resource allocation function *associating every processing stage* Ξ_{jk} *with a* resource allocation request $\mathcal{A}(j,k) \equiv A_{jk}$. *More specifically, each* A_{jk} *is an m-dimensional vector, with its i-th component indicating the number of resource units of resource type* R_i *necessary to support the execution of stage* Ξ_{jk}. *Obviously, in a well-defined RAS,* $A_{jk}(i) \leq C_i$, $\forall j,k,i$.

5 $\mathcal{T}$: $\bigcup_{j=1}^{n} \mathcal{S}_j \rightarrow \mathcal{D}$ *is the* timing function, *corresponding to each processing stage* Ξ_{jk} *a distribution* D_{jk} *that characterizes the statistics of the processing time* t_{jk}, *experienced during the execution of stage* Ξ_{jk}.

Furthermore, $|\Phi| \equiv |\mathcal{R}| + |\bigcup_{j=1}^{n} \mathcal{S}_j| + \sum_{i=1}^{m} C_i$ *will be referred to as the* size *of* Φ.

Some remarks are necessary in order to further clarify the content of Definition 1. First of all, the concept of $\mathcal{G}_j$ introduced in item (3.b) is rather ill-defined. The characterization of the implied data structure can be further concretized through the employment of some additional semantics that allow the effective representation of sequencing specifications. In general, there is a trade-off between the modelling power of the adopted semantics and the extent to which they can facilitate the analysis of the system behavior. Because of this effect, in this book we shall employ a number of representations for expressing the process structure and behavior, always trying to maintain the adopted representation as simple, and therefore, as analyzable, as possible.

In fact, for more complex process behaviors, the characterization and the correctness analysis of the process-defining logic, expressed by the $< \mathcal{S}_j, \mathcal{G}_j >$ tuple, is an interesting issue in itself, studied by modern workflow management theory (Van der Aalst and Van Hee, 2002). Since, however, the main focus of this book is the administration of the resource allocation taking place during the enactment phase of the various processes, we shall not address it any further. Instead, throughout the subsequent analysis, we shall assume that the various RAS processes are well-defined; in particular, we shall request that the underlying process-defining logic satisfies the following conditions:

CONDITION 1 *Under expedient resource allocation, every activated process instance will terminate in a* finite *number of processing steps.*

CONDITION 2 *Every processing stage* $\Xi_{jk} \in \mathcal{S}_j$ *can be realized by at least one execution sequence supported by* $\mathcal{G}_j$.

CONDITION 3 *The only way in which two distinct activated process instances can interact with each other, is through their potential contest for some of the system resources.*

Condition 1 excludes those pathological situations in which an executing process can entangle itself in an infinite loop. In well-designed applications, a process will not be allowed to run within the system indefinitely. From a representational standpoint, the satisfaction of this assumption allows the modelling of the process-defining logic through an *acyclic* data structure. Condition 2 essentially ensures that the process representation does not introduce redundant processing stages. Finally, Condition 3 applies primarily to complex process flows that involve parallelization, and implies that the logic coordinating the execution of the various process threads does not "confound" enacted sub-processes belonging to different process instantiations.

2.2 A RAS taxonomy

While Definition 1, augmented with Conditions 1, 2 and 3, outlines the general class of process behaviors to be considered in this work, we shall define also a number of more specific RAS classes by restricting the admitted process behaviors. These RAS sub-classes present quite rich behavior in terms of, both, complexity and applicability, in order to merit their independent study. Furthermore, they can be hierarchically organized in terms of the complexity of the supporting process behaviors, and therefore, their introduction offers an effective tool for managing the complexity of the analysis and design problems undertaken in this book.

Technically, these subclasses are obtained by imposing some further structure on (i) the process sequencing logic $\mathcal{G}_j$, and (ii) the resource allocation requests A_{jk} associated with processing stages Ξ_{jk}. Regarding the process-defining logic, we consider the following restrictions:

Linear (LIN-)RAS This is the RAS sub-class where each process type Π_j comprises a single execution sequence that is defined as a *total* ordering $\mathcal{G}_j$ of the set of processing stages $\mathcal{S}_j$.

Disjunctive (DIS-)RAS This is the RAS sub-class where the execution logic of every process type Π_j can be represented by an *acyclic digraph*, $\mathcal{G}_j$, with node set equal to $\mathcal{S}_j$. Every path in $\mathcal{G}_j$ connecting a "source" to a "sink" node corresponds to an execution sequence of Π_j, and therefore, this RAS sub-class models process behaviors that present *routing flexibility*.

Coordinating (COR-)RAS This is the RAS sub-class where the execution logic of every process type Π_j can be represented by a *"fork/join"* acyclic digraph,[7] $\mathcal{G}_j$, with node set equal to $\mathcal{S}_j$. Hence, every process type of this RAS sub-class consists of a single main execution sequence, but this sequence involves a number of *"threads"* running in parallel and synchronizing at various stages of their execution.

---

[7]c.f. (Gershwin, 1994) for a systematic characterization of *"fork/join"* – or, alternatively, *"assembly/disassembly"* – networks.

Complex (CPX-)RAS This is the most general RAS class, that can include any combination of the above behaviors, and it only abides to Conditions 1 to 3 stated above.

Regarding the resource allocation function, we consider the following restrictions:

Single-Unit (SU-)RAS This is the RAS sub-class where every resource allocation request A_{jk} is a *unit* vector; i.e., each processing stage in this RAS class is supported by a single unit from a single resource type.

Single-Type (ST-)RAS This is the RAS sub-class where every resource allocation request A_{jk} is a vector with a *single non-zero* entry; i.e., each processing stage in this RAS class is supported by a single resource type (however, it might engage more than one unit of it).

Conjunctive (CON-)RAS This is the RAS sub-class where resource allocation request vectors A_{jk} can be any arbitrary integer vector such that $0 \leq A_{jk}(i) \leq C_i$, $i = 1, \ldots, m$; i.e., these RAS allow for arbitrary resource allocation requests, as long as they are feasible with respect to the system resource availability.

With the exception of the sub-class pair of DIS-RAS and COR-RAS, each of the above two RAS classifications is linearly ordered, in the sense that each RAS-subclass in them is subsumed by those introduced after it. Furthermore, since the criteria employed in the development of those two classifications are independent from each other, one can define a richer, partially ordered taxonomy, by taking the intersection of RAS class pairs, where each RAS class belongs to one of the originally defined taxonomies; for instance, one can consider LIN-SU-RAS or DIS-CON-RAS. Notice that in this refined taxonomy, LIN-SU-RAS constitutes the most restrictive RAS subclass with the lowest complexity, and CPX-CON-RAS is the most encompassing RAS class considered in this book, and therefore, it presents the highest complexity. We shall return to this taxonomy repeatedly, in an effort to effectively manage the complexity underlying the various problems studied in later parts of this book.

2.3   The RAS Supervisory Control problem

Another point that must be elaborated with respect to the operation of the RAS considered in Definition 1, is the detailed dynamics of the resource allocation and deallocation process. In the following analysis, we consider that a process instance j_j advances from stage Ξ_{jk} to a successor stage $\Xi_{j,k+1}$ only upon being allocated the entire set of resources implied by the resource allocation request $A_{j,k+1}$. The allocation of all these resources takes place simultaneously, and it is only at this point that the process instance j_j releases the resources allocated to it for the execution of processing stage Ξ_{jk}. This *"hold-while-waiting"* assumption is introduced in order to capture the resource allocation dynamics pertaining primarily to resources supporting the physical buffering of the executed process instances. Parts processed in a flexibly automated production system or vehicles in an AGV network are physical entities and they always need to be accommodated somewhere during their sojourn through the system. It must be noticed, however, that, while providing the necessary specificity for the underlying resource allocation dynamics, the aforestated assumptions do not compromise the modelling power of our framework, since one can capture any additional resource allocation dynamics by augmenting the specification of process Π_j. For example, one can model the fact that, at some particular process stage Ξ_{jk}, process Π_j might release (some of) its currently allocated resources before advancing to stage $\Xi_{j.k+1}$, by introducing to the process specification an intermediate process stage Ξ_{jq}, with resource allocation request A_{jq} equal to A_{jk} minus the deallocated resource set.

Yet, the above assumption governing the detailed dynamics of the underlying resource allocation is very important from a control-theoretic standpoint, since it can give rise to a highly undesirable behavior in the context of the considered RAS class. More specifically, the fact that a given process instance will hold its currently allocated resources while awaiting for the allocation of the resources necessary to support its next processing stage, when combined with (i) the exclusive nature of the allocation of the system resources, and (ii) the arbitrary structure of the active process routes, can give rise to *"circular dependencies"* among a subset of activated process instances, in a way that it will stall the further progress of these processes, while driving to zero the utilization of the resources currently allocated to them. In the context of the presented theory, such a situation will be characterized as a *RAS deadlock*. A formal characterization of this concept and of the induced RAS dynamics is contingent upon the particular behavioral modes assumed by the various RAS sub-classes defined in Section 2.2, and it will be provided in later parts of this book.

From a conceptual standpoint, there are three primary strategies for dealing with the problem of the RAS deadlock. The first strategy, known as *prevention*, essentially tries to constrain the RAS *structure* in a way that deadlock will never occur during the system operation. In the context of Definition 1, this can be achieved by constraining the sequencing logic and the resource allocation associated with the various processes in a way that ensures that no circular dependencies among the various active process instances are possible. A typical way to achieve such an effect is by enforcing a *"unidirectional"* process flow with respect to the system resources. Technically, this can be achieved by imposing an ordering on the system resources and requesting that each process obtains its resources in increasing order. In a more practical setting, this is essentially the deadlock resolution strategy adopted in the robotic cell when operated under batch processing, or in a tandem AGV system. In the former case, the applied resource ordering is that induced by the sequence in which resources are used by the jobs in the batch,[8] while in the latter, the resource ordering can be any ordering that follows the sense of motion enforced on the guidepath loops. As mentioned in the discussion of the motivating examples, while such a strategy is easily implemented, it tends to stifle the operational flexibility of the system, and eventually its potential performance.

The second basic strategy for resolving the RAS deadlock problem is *detection & recovery*. According to this strategy, deadlock is allowed to occur, but the system is also equipped with a monitoring mechanism that detects its development and triggers an exception handling procedure that resolves it and returns the system to its deadlock-free operational mode. Apparently, the efficiency of such a method is strongly dependent upon the cost of the applied deadlock resolution procedure. Hence, this approach is currently advocated for database transactional systems, where the removal of a running job from the system and its reintroduction at a later point in time can be achieved with a low operational / computational cost. On the other hand, the adoption of such an operational scheme in a heavily loaded manufacturing system could result in excessive material handling costs and long delays as a considerable number of running jobs would have to be withdrawn to some central storage and reintroduced to the system at a later time.

The last primary strategy available for deadlock resolution is *deadlock avoidance*. This strategy addresses the problem of deadlock by controlling the sequence in which resources are granted to the requesting processes. It uses on-line feedback about the current allocation of the system resources – i.e., the RAS *state* – and the available knowledge about the structure of the process types corresponding to the active process instances, in order to prevent the system from getting into resource allocation patterns from which deadlock is unavoidable. In principle, by making informed decisions about the safety of a resource allocation operation, a deadlock avoidance policy can allow the underlying RAS to prevent the occurrence of deadlock while maintaining maximum operational flexibility. In practice this last statement must be further qualified by the fact that in the considered RAS class, the problem of determining the safety of any given RAS state, that underlies the effective implementation of the *maximally permissive deadlock avoidance policy (DAP)*, is NP-complete (Garey and Johnson, 1979). Yet, this strategy remains an interesting proposition for the considered RAS class from, both, a theoretical and a practical

---

[8]in fact, for such an ordering to be properly defined, and for the resulting operational scheme to remain deadlock-free, it is required that no process instance visits a workstation more than once.

standpoint, and its effective implementation will constitute a major part of this book.

From a theoretical standpoint, the aforementioned problem of deadlock avoidance falls in a particular class of control problems known as *behavioral* or *structural* or *logical* control problems. The main concern in this class of problems is to constrain the system behavior, in terms of the generated event sequences, so that it satisfies a certain set of logical specifications and properties. In the case of RAS deadlock avoidance, the required property is *deadlock-freedom*, or equivalently, the establishment of *nonblocking* behavior, i.e., the ability of every activated process instance to proceed to completion through "normal" system operation, and without the need of any external intervention or exception-handling routine. These types of control problems are also known as *preventive control*, since the applied control logic seeks only to prevent potential events / actions that would result in violation of the behavioral specification. In other words, the main effect resulting from the application of a structural or logical control policy is to filter out from the initial *feasible* behavior of the system – i.e., the behavior supported by its original uncontrolled structure – a behavioral sub-space which will constitute the *admissible* behavior (under the imposed policy).

Typically, the admissible behavior will still provide considerable latitude / choice at each decision point. Hence, an additional level of control is required in the operation of these environments that will seek to *bias* their admissible behavior in order to satisfy some additional performance criteria. This type of control will be characterized as *performance-oriented* control; as we saw in the presentation of the motivational examples, in the case of RAS applications, performance-oriented control boils down to the typical sequencing and scheduling problems addressed in the standard OR literature. Taken together, the aforementioned structural / logical and the performance-oriented control problems, provide a complete characterization of the primary concerns to be addressed in the resource allocation function underlying the operation of most contemporary applications. The resulting combined problem will be collectively characterized as the RAS *supervisory control (SC)* problem. Next we discuss some further important aspects of the RAS SC problem and introduce the basic control architecture that will provide the methodological context for our work.

3.   The proposed SC framework

The RAS SC problem defined in the previous section is *large-scale, (highly) stochastic* and *complex*. Its large-scale nature comes from the fact that we intend to target application contexts involving a large set of resource types, frequently available at non-unit capacity levels, and being able to support a large portfolio of process types at each configuration. The problem stochasticity arises from the fact that in many – in fact, most – cases, the distributions D_{jk} characterizing the processing time of stages Ξ_{jk} will be non-deterministic. Things can be further complicated in this direction, if one considers probabilistic outcomes / yields at each processing stage leading to different subsequent routings, and various other contingencies that might alter the running system configuration, e.g., resource outages. The increased problem complexity is partly due to its large-scale and stochastic nature, but, as we shall show in a later chapter, it is also the result of the discrete and combinatorial nature of the underlying system dynamics.

Taken together, these three problem attributes suggest a set of specification requirements that must be met by any solution approach proposed for the RAS SC problem, and, in this way, they outline the set of viable methodological approaches to it. More specifically, in the light of the above discussion, the adopted methodological framework must possess the following properties:

Scalability, i.e., the proposed SC framework must be effectively implementable to any possible RAS configuration in the class encompassed by Definition 1. This requirement can be translated as the need for effective *complexity management*.

Efficiency, i.e., the aforementioned scalability requirement should not come at a very high cost for the system behavioral flexibility and performance. By focusing on some performance aspects, this requirement introduces a notion of *optimality* in the overall design process.

Robustness, i.e., the proposed framework should account for the random and stochastic elements in the system behavior, and provide the mechanisms to effectively accommodate them.

Next we discuss the practical implications for these three specification requirements. Starting with the last, we notice that the ability to accommodate the operational stochasticity resulting from non-deterministic event timings implies that the adopted solution should be based on *closed-loop, feedback-based control* schemes, and not on approaches based on *open-loop planning*. As a more concrete example, trying to establish conflict-free vehicle routing in an AGV system based on an effort to specify a priori the detailed timing of each vehicle movement in the guidepath network is very unlikely to work, since, pretty soon, the underlying operational stochasticity will render such a plan infeasible. In fact, we believe that, beyond the need for closed-loop, feedback-based control schemes, the aforementioned requirement for robustness necessitates the explicit separation of the RAS structural / logical control from the performance-oriented control. As mentioned in the previous section, the approach proposed in this book tries to "curve out" an admissible behavior from the system, based on logical analysis of the allowed event sequences, and it addresses performance considerations only in the latitude provided by this admissible behavior. It should be noticed that, while primarily motivated by the need for robust behavior, the proposed decomposition of the RAS SC problem to structural and performance-oriented control also contributes to the effective management of the underlying complexity; we shall return to the issue of complexity and its effective management shortly. A last issue raised by the requirement of robustness, when considered with respect to the aforementioned operational contingencies, is that of the policy *reconfigurability*: The proposed framework must be able to support the effective reconfiguration of the applied control logic in the case that the underlying RAS changes its structure through the addition or removal of resources and/or process types. Computational efficiency can also be a concern in the context of policy reconfigurability; in general, the more frequent the addressed contingencies, the more efficient must be the adopted policy reconfiguration procedures.

As mentioned in its definition, the requirement for *efficiency* introduces a notion of optimality in the overall design process. For the structural control problem, efficiency implies the establishment of nonblocking – more generally, logically correct – RAS behavior while providing *maximal permissiveness*. This requirement is consistent with – essentially operationalizes – the notion of *flexibility* associated with the considered application contexts. For the performance-oriented control problem, efficiency implies the *biasing* of the admissible RAS behavior in a way that optimizes some performance index, e.g., maximizing the system throughput, minimizing the expected process cycle times, etc. It turns out that, when formulated in the context of the RAS class of Definition 1, all these optimization problems belong to the class of NP-complete problems (Garey and Johnson, 1979). Hence, in the light of the *scalability* requirement, which is the third part of our specification, achieving strictly optimal behavior and performance is a rather hopeless proposition, for most practical settings. However, providing a systematic characterization for the notion of optimality underlying these control problems is very important, since, as it is shown in the following chapters, it can guide our effort towards the development of approximating near-optimal solutions. More specifically, the study of these formulations can provide important qualitative insights regarding the nature and structure of optimality that can be effectively incorporated in some approximating scheme. In addition, the design of algorithms that can effectively solve small-scale versions of these formulations can provide "experimental" data that can be used for the benchmarking of any other approximating policy or algorithm.

The research program underlying the development of this book is motivated by, and effectively synthesizes in a complete methodological framework, all three specification requirements and their practical implications for the RAS SC problem discussed above. The proposed RAS SC framework is effectively

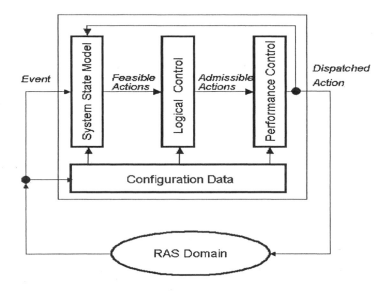

Figure 1.7. The proposed control framework for the RAS SC problem

depicted through the block diagram of Figure 1.7. As indicated in Figure 1.7, the proposed controller is *event-driven*, i.e., the control actions commanded to the underlying RAS can be perceived as the controller responses to the various events taking place in the RAS domain and communicated to the controller through a monitoring function. Hence, the entire control function evolves in a number of *cycles*, with each cycle being triggered by a RAS event communicated to the controller. Conceptually, each cycle consists of three major phases: (i) In Phase I, the controller updates a representation of the RAS *state* so that it represents the RAS status after the occurrence of the communicated event. This representation, combined with the system knowledge about the running RAS configuration, encodes the entire set of *feasible actions* that could be executed by the RAS as a response to the occurring event. (ii) In Phase II, the controller applies the adopted structural control policy in order to filter out from the set of feasible actions identified in Phase I, the set of *admissible actions*, i.e., this set of actions that satisfy some logical specification for the RAS behavior. (iii) Finally, in Phase III, the set of admissible actions is provided to the performance-oriented component of the RAS supervisor in order to select the one that will be communicated eventually to the RAS environment, in a way that observes some performance considerations. In addition to this basic functionality, the RAS controller should be able to respond to the various contingencies taking place in the RAS domain, by (i) appropriately updating the RAS configuration database, and (ii) revising the logical and performance-oriented control logic in order to apply in the emerging RAS configuration. This last function will be collectively characterized as *(re-)configuration management*.

By now we have all the necessary concepts in place, in order to state the basic objective of this book as *the study of the representational frameworks and the analytical methodology that can support the implementation of the RAS SC framework depicted in Figure 1.7, while observing the posed specification requirements of scalability, efficiency and robustness.* At a more conceptual level, we envision this work as contributing towards the eventual development of an *"operating system"* able to support logically correct, efficient and transparent resource allocation in the operational context of the emerging flexibly automated technological applications. Judging from the past experience with the development of the digital computer – clearly the most fully automated complex operational platform produced by modern technology – the availability of a powerful formal theory to support the analysis and design of the underlying system behavior and the resulting performance, is of paramount importance for the effective deployment of the target application concept, and its eventual success and acceptance by the broader community ((Gates, 1995), Chpt 3). The next section provides a more detailed outline of the book contents and contributions.

4. The book organization and an outline of the key results

As it was stated in the previous section, the material of this book seeks to provide the representational frameworks and the analytical methodology for supporting the implementation of the RAS SC framework depicted in Figure 1.7, in a robust, efficient and scalable manner. Hence, the book development is naturally organized around the basic structure and functionality presented by the aforementioned reference framework.

More specifically, Chapter 2 initiates the study of the RAS logical control problem, seeking to develop a formal characterization of the RAS behavior and of the optimal – i.e., maximally permissive – supervisory control policy (SCP) that will guarantee nonblocking operation. In order to enhance the specificity and tractability of the presented results, this first analysis is performed in the context of DIS-CON-RAS. It is shown that the modelling framework of *Finite State Automata (FSA)* (Hopcroft and Ullman, 1979) provides a natural and conceptually straightforward characterization of the DIS-CON-RAS behavior and of the corresponding maximally permissive non-blocking supervisor. It is also established, however, that the problem of implementing the optimal SCP, on a given RAS configuration, belongs to the notorious class of NP-Hard problems (Garey and Johnson, 1979), even for the simpler case of LIN-SU-RAS. In the light of this last result, the concept of *Polynomial-Kernel (PK-)* SCP is proposed, as a polynomial-complexity approximation to the optimal

SCP: PK-SCP's are expected to provide polynomial-complexity nonblocking supervision of the underlying RAS, while maintaining a considerable portion of its operational flexibility. The last part of Chapter 2 considers the problems of (i) establishing optimal nonblocking supervision for DIS-CON-RAS where some processes can advance or determine their routing in an uncontrollable manner, and (ii) the accommodation of operational contingencies, like resource failures, in the applied control logic.

Chapter 3 presents a series of results establishing that, in spite of the negative complexity result developed in Chapter 2, for a large part of the SU-RAS class, optimal nonblocking supervision is of polynomial complexity with respect to the underlying RAS size. These results are of very practical significance, as they apply to many of the application contexts that originated the problem of the RAS nonblocking supervision – e.g., the buffer space allocation in flexibly automated production systems – and they provide guidelines for the design of these environments in a way that supports effective implementation and management of the sought operational flexibility. From a theoretical standpoint, these results lead to a more profound understanding of the qualitative dynamics underlying the operation of the considered RAS classes, and they give rise to a new set of methodological tools to be employed in the investigation and characterization of their behavioral patterns and properties.

Chapter 4 undertakes the development of PK-SCP's for RAS classes that do not admit optimal nonblocking supervision of polynomial complexity with respect to the RAS size. In this chapter, the problem is primarily investigated in the more confined, and therefore, simpler context of LIN-SU-RAS. The restriction of this initial discussion to the LIN-SU-RAS class is motivated by the desire to provide a smooth and clear introduction of the reader to the relevant concepts and theory, but also, by the considerable significance of this class in terms of applications, and by the richness of the relevant theory. Indeed, LIN-SU-RAS is the most extensively studied RAS class in the current literature, while the relevant simplicity of the supported behavior allows the employment of intuitive representational tools and arguments that are not applicable to the behavioral context of the more complex RAS classes. The first part of the chapter provides a number of PK-SCP's that are appropriate for the LIN-SU-RAS class, and it also discusses some techniques that can effectively enhance the permissiveness of these SCP's when applied on any given LIN-SU-RAS configuration. The second part of the chapter extends the aforementioned results in order to develop a behavioral model and a nonblocking supervisor for the RAS class characterizing the logical behavior of the zone-controlled AGV system that was introduced in the second example of Section 1. The last part of Chapter 4 surveys a number of results that employ and extend the aforementioned PK-SCP's for LIN-SU-RAS, in order to accommodate various operational contingencies arising in these RAS; the presented results constitute a concretization, in the considered operational context, of the more general and abstract ideas introduced in Chapter 2.

Chapter 5 considers the problem of synthesizing PK-SCP's for RAS that exhibit more complex process behavior. The development of this chapter is carried out in the modelling framework of *Petri nets (PN)* (Murata, 1989), that facilitates the effective representation of the behavioral complexity of the considered RAS classes, and provides specificity and rigor in the presented discussion. Hence, the first part of this chapter provides a systematic characterization of the structure and behavior of the considered RAS classes in the PN modelling framework. Subsequently, a PN-based characterization of the RAS deadlock is developed, and it is shown to be instrumental for assessing the correctness of PK-SCP's synthesized for these broader RAS classes. It is also shown that the results of Chapter 4, concerning the nonblocking supervision of the class of LIN-SU-RAS, can provide the stepping stones for synthesizing effective SCP's for the more complex RAS classes considered in this chapter. At the same time, the aforementioned PN-based structural characterization of the RAS deadlock, and the ensuing approaches to SCP correctness verification, are proven to be very useful for the further enhancement of the flexibility of the PK-SCP's developed in Chapter 4, as they enable the identification of more permissive instantiations of these policies. The chapter concludes with an extension of the SCP's developed in its earlier parts, so that they can effectively accommodate the RAS uncontrollability characterized in Chapter 2.

Chapter 6 turns to the problem of optimizing the performance of RAS that are controlled by the logical control policies developed in Chapters 2-5. This problem essentially corresponds to the scheduling problem, that has been typically addressed by the Operations Research literature (Hillier and Lieberman, 2002), formulated, however, on the behavioral subspace that is admitted by the applied logical control policy, instead of the original behavioral space of the underlying RAS. Therefore, all the heuristical solutions that have been adopted, in the form of *dispatching rules*, as practical solutions to the various formulations of the real-time scheduling problem, in the face of their NP-Hard nature, can also apply to the problem under consideration. Chapter 6 starts with this basic observation, but subsequently, it seeks to provide a formal characterization of the optimal scheduling policy for logically RAS, and to outline potential approaches that can lead to the effective and efficient approximation of this policy while maintaining polynomial implementational complexity with respect to the underlying RAS size. From a methodological standpoint, *Continuous-Time Markov Decision Processes (CTMDP)* (Puterman, 1994) is the framework employed for the first task, while the novel area of *Neuro-Dynamic Programming (NDP)* (Bertsekas and Tsitsiklis, 1996) is contemplated as a potential approach for the second. A third part of the chapter also demonstrates the ability of the aforementioned MDP modelling framework to rationalize any decisions pertaining to the configuration of the underlying RAS structure and the applied SCP, by enabling the characterization of their impact on the system performance. All of the above problems are addressed in the class of DIS-CON-RAS, and under the particular performance objective of maximizing the long-run system throughput; however, the entire development has a prototypical character, and the derived results can be easily extended to other problem formulations.

Chapter 7 summarizes the key results and contributions of this book, and it identifies a series of additional problems that must be addressed in order to complete the theory and provide the computational capability that is necessary for the full deployment of the framework of Figure 1.7.

From a methodological standpoint, the framework of Figure 1.7, and many of the results developed in the rest of this book, constitute an integrated and systematic application to the RAS supervisory control problem of ideas and techniques developed in the control-theoretic area of *Discrete Event Systems (DES)* (Cassandras and Lafortune, 1999). More specifically, the aforementioned representational frameworks of FSA's, PN's, and MDP's, that are employed for the formal characterization of the logical and/or performance-oriented control problems addressed in this work, as well as the notions of optimality associated with the solutions of these problems, are borrowed directly from the DES theory. However, the material of this book complements these characterizations with a systematic *approximation theory* that enables the effective deployment of the original DES framework in the large-scale operational contexts underlying the target RAS applications. Another methodological contribution of this book is that it employs the aforementioned representational frameworks in a complementary and synergistic fashion, demonstrating, in the process, the analytical and computational gains that can be achieved through such an integrated approach. Finally, to the best of the author's knowledge, the current results concerning the interaction and integration of the logical and performance-oriented control policies that are typically needed for the effective deployment of any contemporary flexibly automated technological application, are very limited. Hence, the results of Chapter 6 can be claimed as the first effort to address extensively and systematically this problem.

5. Historical and bibliographical notes

This section complements the material developed in this chapter, by pointing to some further readings that can support the concepts and ideas introduced in it, and by providing some contextual and historical comments and notes. As it was mentioned in the opening paragraph of this chapter, the concept of resource allocation is very broad, and already extensively used in the academic literature. One can argue that the entire area of queuing theory (Medhi, 1991) – a well-defined area in Operations Research (Hillier and Lieberman, 2002) – essentially deals with the problem of resource allocation. However, with minimal exceptions, most of the modelling abstractions utilized by this area target primarily performance considerations, and they lack the specificity necessary to capture the operational aspects pertaining to the logical issues of the system behavior. For example, the modelling framework of *multi-class queueing networks (MCQN's)* (Kumar, 1994b), one of the broadest and most elegant analytical tools offered by contemporary queueing theory, considers only the contest of the active processes for the finite processing capacity of the system workstations, while completely ignoring the finiteness of the system buffers and the corresponding blocking effects. As it will be shown in Chapter 6, due to the ignorance of these blocking phenomena, results and insights based on MCQN's are not immediately transferrable to the RAS environment and the SC problem considered in this work. On the other hand, queueing networks with blocking (Perros, 1994) have tried to explicitly address some of these blocking effects, but the developed models, in their effort to maintain analytical tractability, tend to severely restrict the admissible RAS structure and behaviors.

To emphasize the preoccupation of the applied OR-related literature with the performance related considerations, and the almost complete ignorance of behavior-related problems, it is worth-noticing that the typical implementation of the *hierarchical control* (Gershwin, 1989) framework in production systems – which incidentally is the most widely accepted framework for the overall production planning and control problem – recognizes (i) strategic, (ii) tactical and (iii) operational-level decisions, all of which are abstracted as an optimization problem targeting some performance objective, but fail to effectively model the operational details governing the actual system behavior (Askin and Goldberg, 2002). Such an approach has been facilitated in the past by the presence of the human operator who has actively intervened in order to establish the viability of the production plans / solutions derived through the aforementioned formulations. As these systems move to more extensively automated modes of operation, the inadequacy of the current practice to support the target behaviors and performance becomes increasingly apparent; for some publications registering this issue, the reader is referred to (Joshi et al., 1995; Suarez et al., 1997). One particular set of manufacturing-related policies that have some stronger affinity to this work, in the sense that they are trying to control explicitly the distribution of the system workload among the different workstations, are those based on the KANBAN concept (Hopp and Spearman, 1996). However, these policies are still motivated by performance rather than behavioral considerations – after all, they target primarily *flowlines*, i.e., unidirectional material flows – and the analytical results on them are rather limited (Glasserman and Yao, 1994; Perkins and Kumar, 1995; Di Mascolo et al., 1996; Dallery and Liberopoulos, 2000).

Yet, the last 10-15 years have been characterized by considerable effort that, motivated by the aforementioned experiences, tries to systematically model, analyze and control the behavioral aspects of the resource allocation underlying the operation of flexibly automated environments, with particular emphasis on manufacturing-related applications. Some seminal visionary works characterizing the emerging needs are those published, for instance, in (Naylor and Volz, 1987; Lawley et al., 1997a; Joshi et al., 1995), while the books of (Desrochers, 1990; Viswanadham and Narahari, 1992; Zhou and DiCesare, 1993) provide a comprehensive view of the relevant activity and effectively summarize the spirit of the undertaken analysis and the derived results up to their publication time. This effort has recently escalated to a large number of publications appearing in journals like the IEEE Trans. on Automatic Control, Robotics & Automation, Systems, Man & Cybernetics, the IIE Trans., the International Journal of Production Research, the International Journal of FMS, and other manufacturing-related journals.

One particular development that supported the aforementioned research activity by providing it with a stronger methodological context, identity and visibility, was the emergence of *Logical Supervisory Control* theory as a distinct area of modern control theory. This field was founded, to a large extent, on the seminal work by Peter Ramadge and Murray Wonham (Ramadge and Wonham, 1989), and employs the formal framework of *(finite state) automata* and *formal languages* (Hopcroft and Ullman, 1979) in order to rigorously characterize the logical aspects of the system behavior and the relevant control problems. Currently, the main contributions of the field include the rigorous characterization of the concepts and issues pertaining to behavioral control, and the systematic assessment of the effective computability of the formulated problems. On the other hand, most of the proposed solution approaches suffer from very high (non-polynomial) computational complexity, and therefore, they lack scalability and applicability to most "real-world" application contexts. For a systematic introduction to this area and its most classical, by now, results, the reader is referred to (Cassandras and Lafortune, 1999; Kumar and Garg, 1995; Moody and Antsaklis, 1998). Some interesting complexity and computability results are reported in (Gohari and Wonham, 2000; Rohloff and Lafortune, 2003). The largest body of archival publications in this area can be found in IEEE Trans. on Automatic Control, the Intl. Journal of Discrete Event Systems, and the SIAM Journal of Optimization & Control.

Another area of modern control theory that is relevant to the RAS SC framework presented in Section 3, and which actually encompasses the aforementioned Ramadge & Wonham's SC framework, is that of *Discrete Event (Dynamical) Systems (DES)*. According to (Cassandras and Lafortune, 1999), which is the main introductory reference in the area, a DES is a discrete-state, event-driven system, i.e., its state evolution depends entirely on the occurrence of asynchronous discrete events over time. The event-driven nature of the DES dynamics differentiates them from the more classical domain of time-driven systems, and gives rise to a new set of models and techniques for their representation, analysis and control. Specifically, (Cassandras and Lafortune, 1999) recognizes three levels of abstraction modelling DES behavior, which are formalized through the concepts of *untimed, timed* and *stochastic timed languages*. Among these three levels of representation, the untimed language (or more simply, language) provides a formal expression of the "logical behavior" of the underlying DES, by providing the set of all possible event sequences that could happen in the considered system. In this modelling context, the actual or the expected timing of the various events is not considered, and the emphasis is placed on the structural properties of the generated system behavior. The relevant analysis is further facilitated by the introduction of additional *discrete event modelling formalisms*, that allow the DES language representation in a manner that highlights structural information about the system behavior, and renders convenient its manipulation during the analysis and the controller synthesis phases. The two main formalisms employed in the logical modelling of the DES behavior are those of (i) *(finite state) automata* (Hopcroft and Ullman, 1979), and (ii) *Petri nets* (Murata, 1989). The main formalisms supporting the modelling of the timed system behavior, are *timed automata* (Brandin and Wonham, 1994) and *stochastic processes* (Gallager, 1996). We shall return to all these models at different points of the book development, since, as it was pointed out in Section 4, the RAS SC control framework of Figure 1.7 constitutes an implementation of the general DES control framework in the particular context of the RAS SC problem.

Regarding the RAS deadlock avoidance problem, that, as it was mentioned in Section 2, will be a central issue in this book, we make the following remarks: The problem was originally identified and studied in the context of Computer System Engineering, back in the late 1960s and early 1970s. Some seminal works coming from that area are reported in (Dijkstra, 1965; Havender, 1968; Habermann, 1969; Coffman et al., 1971; Holt, 1972). Characteristically, *Banker's* algorithm, perhaps the most well known deadlock avoidance policy, was developed by Dijkstra in (Dijkstra, 1965). Subsequently, the works of (Araki et al., 1977; Gold, 1978) studied systematically the problem complexity for a number of RAS configurations. Since the early 1990's, the problem has been revisited in the context of the sequential resource allocation arising in contemporary FMS and other technological applications. These results, including the results appearing in this book, have been primarily published in IEEE Trans. on Robotics & Automation, Automatic Control, and Systems, Man & Cybernetics, IIE Trans. and the Intl. Journal of FMS.

Finally, a more extensive discussion on the technological applications introduced in the motivational examples of Section 1, can be found in the following references: Flexibly Automated Production Systems – (Groover, 1996), AGV systems – (Ganesharajah et al., 1998), Semiconductor Manufacturing Material Handling Technologies and Practices – (Mackulak et al., 2002), Workflow Management Systems – (Lawrence, 1997; Van der Aalst and Van Hee, 2002), Internet-based Computing – (Marinescu, 2002).

Chapter 2

LOGICAL CONTROL OF
DISJUNCTIVE / CONJUNCTIVE
RESOURCE ALLOCATION SYSTEMS

In this chapter we undertake the systematic investigation of the RAS logical control problem, that was introduced in Sections 2 and 3 of Chapter 1. However, at this first stage, our analysis will be confined to the behavioral context of the *Disjunctive / Conjunctive (DIS-CON-)RAS* sub-class, that was defined in the taxonomy of Section 2.2, in Chapter 1. We remind the reader that this RAS sub-class allows for arbitrary resource allocation requests and process routing flexibility, but it does not allow for process parallelization. Hence, at every point in time, every active process instance in the resource allocation system constitutes a single atomic entity, executing one of its processing stages. The extension of the relevant theory in order to encompass more complex RAS behaviors, involving (sub-)process coordination, requires a more sophisticated modelling framework and more complicated analysis tools, and it is deferred to Chapter 5. In the sequel, any invocation of the RAS concept should be considered in the context of the DIS-CON-RAS sub-class, unless otherwise specified.

According to the discussion of Section 2.2, in Chapter 1, the class of DIS-CON-RAS is obtained from Definition 1 by requiring that, for each process type Π_j, the corresponding data structure $\mathcal{G}_j$ is an *acyclic graph* with node set equal to the set of processing stages, $\mathcal{S}_j$. Each edge (Ξ_{jk}, Ξ_{jq}) of this graph implies that processing stage Ξ_{jq} can be an *immediate successor* stage for processing stage Ξ_{jk} (or, equivalently, that processing stage Ξ_{jk} can be an *immediate predecessor* stage for processing stage Ξ_{jq}). Furthermore, we shall use the notation $\mathcal{S}_j^{\nearrow}$ (resp., $\mathcal{S}_j^{\searrow}$) in order to refer to the set of stages that correspond to *source* (resp., *sink*) nodes of $\mathcal{G}_j$. Every path from a "source" stage $\Xi_{jk} \in \mathcal{S}_j^{\nearrow}$ to a "sink" stage $\Xi_{jq} \in \mathcal{S}_j^{\searrow}$ corresponds to a potential *"process plan"* for process type Π_j.

For this simpler class of systems, we employ the formal framework of *Finite State Automata (FSA)* (Hopcroft and Ullman, 1979) in order to provide a natural yet rigorous characterization of the RAS behavior, and of the concepts of *safety* and *maximally permissive nonblocking supervision*, that were informally introduced in the discussion of Chapter 1. The rigorous characterization of the considered logical control problem subsequently enables the systematic study of its computational complexity. It turns out that the problem belongs to the notorious class of NP-hard problems (Garey and Johnson, 1979), and this result establishes a trade-off between the permissiveness of the proposed non-blocking SC policies and their computational tractability. Hence, in the third part of this chapter we outline a general methodology for resolving this trade-off; detailed implementation of the presented ideas in the context of various RAS sub-classes will be provided in Chapters 4 and 5. The chapter concludes with the discussion of some variations of the RAS logical control problem that incorporate (i) potentially uncontrollable elements in the RAS behavior, and (ii) operational contingencies that will necessitate the re-design of the applied supervisory control policy.

1. Finite State Automaton-based modelling of the RAS behavior

As it was explained in the introductory chapter, logical control focuses on the *logical* or *qualitative* properties of the RAS dynamics, and it is concerned with the *logical sequencing* of the various resource allocation - related events taking place in the system. Furthermore, it was argued in that chapter that any approach trying to enforce a particular event sequence in the system behavior by controlling the exact timing of the occurrence of the various events of interest, would be too *brittle* in the considered application contexts, due to the stochasticity of the sojourn times associated with the various processing stages. Hence, this part of our analysis will ignore time completely, and it will employ only *untimed* models providing qualitative yet formal characterizations of the RAS behavior.

1.1 Finite State Automata: Basic Concepts and Definitions

Among the class of qualitative behavioral models employed by Discrete Event System theory, the most straightforward, and, probably, the most widely used, is the *Finite State Automaton (FSA)* (Hopcroft and Ullman, 1979; Cassandras and Lafortune, 1999). A formal definition of this model is as follows:

DEFINITION 2 *(Cassandras and Lafortune, 1999) A* (Deterministic) Finite State Automaton (FSA) *G is a 6-tuple*

$$G = < E, S, f, \Gamma, s_0, S_m >$$

where

- *E is a finite set, called the* event set *of the automaton;*

- *S is a finite set, called the* state set *of the automaton;*

- $f : S \times E \rightarrow S$, *is the* state transition function, *i.e.,* $\forall s \in S, \forall e \in E$, $f(s, e) = s'$ *means that there is a transition from state s to state s' that is triggered by event e; in general, f is a partial function on its domain, i.e., certain events cannot occur in state s;*

- $\Gamma : S \rightarrow 2^E$ *is the* feasible event function, *i.e.,* $\forall s \in S$, $\Gamma(s)$ *denotes the set of all events e for which $f(s, e)$ is defined;*

- $s_0 \in S$ *is the* initial state *of the automaton;*

- $S_m \subseteq S$ *is the set of* marked states.

The transitional structure expressed by the FSA model can be visualized by a labelled graph $\mathcal{G}$; this graph is known as the *state transition diagram (STD)* of the FSA, and its node set corresponds to the state set S of the automaton, its edge set is defined by the state transition function, and its edge label set corresponds to the event set E of the automaton.

It is obvious from the above definitions that the FSA and its corresponding STD can be perceived as a complete map for the behavioral evolution of the modelled system. This effect can be formalized as follows:

DEFINITION 3 *(Cassandras and Lafortune, 1999) Consider an FSA $G =< E, S, f, \Gamma, s_0, S_m >$ and let E^* denote the set containing all the* finite-length *sequences that can be generated from E, including the empty sequence ϵ.*

1 The FSA state transition function f is naturally extended to $S \times E^*$ as follows:

$$\forall s \in S, \quad f(s, \epsilon) \equiv s \qquad (2.1)$$

$$\forall s \in S, \forall u \in E^*, \forall e \in E, \quad f(s, ue) \equiv f(f(s, u), e) \qquad (2.2)$$

2 The language $\mathcal{L}(G)$ generated *by G is defined by*

$$\mathcal{L}(G) \equiv \{u \in E^* : f(s_0, u)!\}^1 \qquad (2.3)$$

---

[1]i.e., $f(s_0, u)$ involves only transitions corresponding to feasible events, and therefore, it is well-defined

3 *The* language $\mathcal{L}_m(G)$ marked *by G is defined by*

$$\mathcal{L}_m(G) \equiv \{u \in \mathcal{L}(G) : f(s_0, u) \in S_m\} \tag{2.4}$$

In the STD context, $\mathcal{L}(G)$ can be described as the set of all event sequences $u \in E^*$ that can be traced on any path, not necessarily simple, starting from the initial state s_0. $\mathcal{L}_m(G)$ is used to model event sequences that correspond to the achievement of some "milestone" in the system behavior.

1.2   FSA-based modelling of the RAS behavior

The formalism and concepts introduced in the previous section, provide all the necessary mathematical apparatus for developing an FSA-based characterization of the RAS behavior. We proceed to this characterization by providing first the formal definition of the RAS *state* that will be employed in the logical analysis of its behavior.

DEFINITION 4 *Consider a RAS $\Phi = < \mathcal{R}, C, \mathcal{P}, \mathcal{A}, \mathcal{T} >$. For the purposes of logical analysis, its* state *$s(t)$ at time t is defined as a vector of dimensionality $D = \sum_{j=1}^{n} l(j)$ – i.e., equal to the total number of distinct processing stages in the system – and with components $s(q;t)$, $q = 1, \ldots, D$, being in one-to-one correspondence with the RAS processing stages, Ξ_{jk}, $j = 1, \ldots, n$, $k = 1, \ldots, l(j)$. Furthermore, component $s(q(j, k); t)$, corresponding to processing stage Ξ_{jk}, indicates the number of process instances executing stage Ξ_{jk} at time t.*

A natural way to define the correspondence between the state components and the RAS processing stages is by setting $q(j, k) = k + \sum_{r=1}^{j-1} l(j)$; this will be the mapping assumed in the following, unless otherwise stated. Also, to simplify the notation, in the following discussion we omit the dependence of state s on time t.

Notice that the information contained in the RAS state is sufficient for the determination of the distribution of the resource units to the various process stages, as well as of the *slack* (or *idle*) resource capacity in the system. In particular, we define the *slack* capacity, $\delta_i(s)$, of resource R_i at sate s, by

$$\delta_i(s) \equiv C_i - \sum_{q=1}^{D} s(q(j, k)) \cdot A_{jk}(i) \tag{2.5}$$

Then, the set S of *feasible* resource allocation states for the considered RAS is defined by

$$S \equiv \{s \in (Z_0^+)^D : \delta_i(s) \geq 0, \ \forall i = 1, \ldots, m\} \tag{2.6}$$

The finiteness of the resource capacities implies that $card(S) \equiv |S| < \infty$. However, in general, $|S|$ will be a *super-polynomial* function of the RAS size; in the particular case where each process stage requires a single unit from a single resource type, $|S| = \prod_{i=1}^{m} \frac{(C_i + |\mathcal{S}(R_i)|)!}{C_i! |\mathcal{S}(R_i)|!}$, where C_i denotes the capacity of resource R_i and $\mathcal{S}(R_i)$ denotes the set of processing stages requesting resource R_i for their execution.

The set of *events*, E, that can change the system state, comprises: (i) the events e_{jk}^{l}, $j = 1, \ldots, n$, $k \in \{1, \ldots, l(j)\} : \Xi_{jk} \in \mathcal{S}_{j}^{\nearrow}\}$, corresponding to the *loading* of a new instance of process type Π_j into the system, that is to follow a process plan starting with stage Ξ_{jk}, (ii) the events e_{jkh}^{a}, $j = 1, \ldots, n$, $k = 1, \ldots, l(j)$, $h \in \{q : (\Xi_{jk}, \Xi_{jq}) \in \mathcal{G}_j\}$, corresponding to *advancement* of a process instance executing stage Ξ_{jk} to a successor stage Ξ_{jh}, and (iii) the events e_{jk}^{u}, $j = 1, \ldots, n$, $k \in \{1, \ldots, l(j)\} : \Xi_{jk} \in \mathcal{S}_{j}^{\searrow}\}$, corresponding to the *unloading* of a finished process instance of type Π_j, whose last processing stage was stage $\Xi_{jk} \in \mathcal{S}_{j}^{\searrow}$. Without loss of generality, it is assumed that, during a single state transition, only one of these events can take place. The resulting transition, however, is *feasible* only if the additionally requested set of resources can be obtained from the system slack capacity; i.e., for each state s, the set of *feasible* events, $\Gamma(s)$, contains only those events that abide to the resource allocation dynamics described in Section 2 of Chapter 1. Based on the above, the *state transition function* f and the *feasible event function* Γ for this automaton, are formally defined as follows:

$$\forall s \in S, \forall e \in E,$$

$$f(s, e) \equiv \begin{cases} s + \mathbf{1}_{q(j,k)} & \text{if } e \equiv e_{jk}^{l} \text{ and } s + \mathbf{1}_{q(j,k)} \in S \\ s - \mathbf{1}_{q(j,k)} + \mathbf{1}_{q(j,h)} & \text{if } e \equiv e_{jkh}^{a} \text{ and } s - \mathbf{1}_{q(j,k)} + \mathbf{1}_{q(j,h)} \in S \\ s - \mathbf{1}_{q(j,k)} & \text{if } e \equiv e_{jk}^{u} \text{ and } s - \mathbf{1}_{q(j,k)} \in S \\ \text{undefined} & \text{otherwise} \end{cases}$$

$$(2.7)$$

$$\forall s,$$

$$\Gamma(s) \equiv \{e \in E : f(s, e)!\} \tag{2.8}$$

A natural definition for the *initial* state s_0 is

$$s_0 \equiv \mathbf{0} \tag{2.9}$$

i.e., the state in which the system is idle and empty of any process instances. Since the main logical concern in this book is the establishment of deadlock-free execution of all activated process instances, we define S_m as

$$S_m \equiv \{s_0\} \tag{2.10}$$

Hence, the *marked language* $\mathcal{L}_m$ of this automaton corresponds to *"complete (processing) runs"*.

1.3   Example

To exemplify the FSA-based modelling of the RAS behavior, consider a small RAS where the system resource set is $\mathcal{R} = \{R_1, R_2, R_3\}$, with $C_i = 1$, $i = 1, 2, 3$. The process types supported by the system possess the linear structure described by the following resource allocation sequences:

$$\Pi_1 = < \begin{pmatrix} 1 \\ 0 \\ 0 \end{pmatrix}, \begin{pmatrix} 0 \\ 1 \\ 0 \end{pmatrix}, \begin{pmatrix} 0 \\ 0 \\ 1 \end{pmatrix} >$$

$$\Pi_2 = < \begin{pmatrix} 0 \\ 0 \\ 1 \end{pmatrix}, \begin{pmatrix} 0 \\ 1 \\ 0 \end{pmatrix}, \begin{pmatrix} 1 \\ 0 \\ 0 \end{pmatrix} >$$

The state of this RAS has six components, corresponding to each of the six processing stages; in particular, state $s_0 = (0\ 0\ 0\ 0\ 0\ 0)^T$ denotes the initial empty state. Table 2.1 enumerates the RAS state transition function. As it can be seen in this table, the considered RAS can present 27 distinct allocation states, i.e., $|S| = 27$, with the state signatures running from 0 to 26. Figure 2.1 provides the corresponding *State Transition Diagram (STD)*.

It is interesting to notice that the RAS considered in this example and the derived FSA provide an effective characterization of the untimed / logical dynamics of the small robotic cell considered in the introductory example of Chapter 1. Indeed, the reader can convince herself that the allocation of the robotic manipulator to a process instance is not of particular interest in the logical analysis of the resource allocation taking place in that cell, since this allocation only facilitates the process transfers among the various workstations, and will remain problem-free as long as the allocation of the workstation capacity is properly managed. As a more general principle, developing and exploiting this type of insights in the problem formulation and analysis is very important since it can lead to a pertinent economical representation of the problem under consideration, and thus, it constitutes a first step towards managing its complexity.
◇

2.   State Reachability, Safety, and Nonblocking Supervision

In this section, we use the STD of the example RAS presented in Section 1.3, in order to motivate and define some additional concepts that are necessary for the eventual formal definition of the RAS nonblocking SC problem.

Table 2.1. Example: The RAS State Transition Function

s_i	State Vector	Succ. States	s_i	State Vector	Succ. States
0	0 0 0 0 0 0	1, 2	14	0 0 0 1 1 1	8
1	1 0 0 0 0 0	3, 15	15	1 0 0 1 0 0	16, 17
2	0 0 0 1 0 0	4, 15	16	0 1 0 1 0 0	18
3	0 1 0 0 0 0	5, 6, 16	17	1 0 0 0 1 0	19
4	0 0 0 0 1 0	7, 8, 17	18	1 1 0 1 0 0	
5	1 1 0 0 0 0	9, 18	19	1 0 0 1 1 0	
6	0 0 1 0 0 0	0, 9	20	0 0 1 0 0 1	6, 7
7	0 0 0 0 0 1	0, 10	21	0 1 0 1 0 1	16
8	0 0 0 1 1 0	10, 19	22	1 0 1 0 1 0	17
9	1 0 1 0 0 0	1, 11	23	0 1 0 0 0 1	20, 21
10	0 0 0 1 0 1	2, 12	24	0 0 1 0 1 0	20, 22
11	0 1 1 0 0 0	3, 13	25	0 1 1 0 0 1	11, 23
12	0 0 0 0 1 1	4, 14	26	0 0 1 0 1 1	12, 24
13	1 1 1 0 0 0	5			

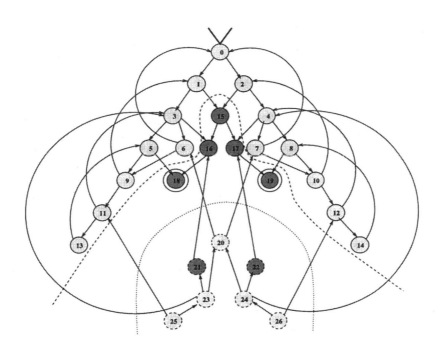

Figure 2.1. Example: The RAS State Transition Diagram

2.1   State Reachability and Safety

As indicated in the introduction of the FSA model, a *directed path* in the STD of Figure 2.1 represents a feasible sequence of events that can take place in the behavior of the considered RAS. We are mainly interested in paths that start and finish at the empty state s_0. Notice that there is a subset of nodes, shown as dashed nodes in the depicted STD, that cannot be reached from state s_0 through any directed path. This implies that when the system is started from empty state, and operated according to the resource allocation dynamics expressed by Equation 2.7, the resource allocation states represented by the dashed nodes will never occur. These states will be referred to as *unreachable*. The remaining states are feasible states under normal operation and they will be called *reachable* states. Formally,

DEFINITION 5 *State s' is* reachable *from state s, denoted by $s' \leftarrow s$, or equivalently, $s \rightarrow s'$, if and only if (iff) there exists a sequence of events that can bring the system from state s to state s'. In the FSA notation,*

$$\forall s, s' \in S, \ s' \leftarrow s \Longleftrightarrow s \rightarrow s' \Longleftrightarrow \exists u \in E^* : f(s, u) = s'$$

Furthermore, a state $s \in S$ will be called a reachable state, *iff $s \leftarrow s_0$.*

The set of reachable states will be denoted by S_r and the set of unreachable states will be denoted by $S_{\bar{r}}$; obviously $S_{\bar{r}} = S \backslash S_r$. The STD subgraph consisting of the reachable states S_r and the arcs emanating from them is called the *reachability graph* of the FSA.

Another important classification of the STD nodes / states results from the following observation: there are states from which the empty state s_0 is reachable by following a directed path of the STD, and states for which this is not possible. In the STD of Figure 2.1, the former are lightly shaded while the latter are heavily shaded. If the RAS enters any of the heavily shaded states it will never be able, under normal operation, to complete all running jobs, i.e. become idle and empty. For this reason, the heavily shaded states are characterized as *unsafe*, while the lightly shaded states, which provide accessibility to state s_0, are characterized as *safe*. Formally,

DEFINITION 6 *State s is a* safe *state iff state s_0 is reachable from state s. A state which is not safe will be called an* unsafe *state. In the FSA notation,*

$$\forall s \in S, \ \mathtt{safe}(s) \Longleftrightarrow \exists u \in E^* : f(s, u) = s_0 \Longleftrightarrow s_0 \leftarrow s$$

The set of safe states is denoted by $S_s \subseteq S$ and the set of unsafe states is denoted by $S_{\bar{s}}$. Again, it holds that $S_{\bar{s}} = S \backslash S_s$. Furthermore, we extend the characterization of safety to RAS transitions emanating from safe states, by characterizing them as safe if they result in a safe state; mathematically,

$$\forall s \in S_s, \ \forall e \in \Gamma(s), \ \texttt{safe}(e|s) \Longleftrightarrow f(s,e) \in S_s$$

Finally, we denote the intersection of any two classes resulting from the previous two classifications by S_{xy} where $x \in \{r, \bar{r}\}$, and $y \in \{s, \bar{s}\}$.

2.2 Safety and the RAS Deadlock

In the class of DIS-CON-RAS, a *deadlock* state is a state where there exists a subset of activated processes, such that every process in this subset requests resources for its advancement currently held by some other process(-es) in the set. More formally,

DEFINITION 7 *Given a RAS state s, define the* apparent slack *of resource R_i with respect to (w.r.t.) a set of active processes, $\mathcal{DP}$, to be equal to the capacity of R_i, C_i, minus the number of its units allocated to processes in $\mathcal{DP}$.*

Then, state s is a (partial) deadlock, *if there exists a set of active processes, $\mathcal{DP}$, with every process j_j in it requesting a number of units from a resource R_{i_j} that is larger than the apparent slack of R_{i_j} w.r.t. set $\mathcal{DP}$.*

Some deadlock states in the STD of Figure 2.1, are states s_{16}, s_{17}, s_{18} and s_{19}. The class of *deadlock* states for a given RAS configuration will be denoted by S_d.

Deadlocks are the natural reason for the existence of the unsafe states in the RAS operation. This is established by the following two propositions, originally proven in (Reveliotis, 1996):

PROPOSITION 1 *A RAS deadlock state is an unsafe state, i.e., $S_d \subseteq S_{\bar{s}}$.*

Proof: Consider a state s satisfying the deadlock characterization of Definition 7. According to the logic expressed by Equation 2.7, a transition corresponding to a state-event pair is feasible *iff* the corresponding resource request can be met by the system slack capacity. Furthermore, resource units become idle only when an occupying process instance proceeds to its next processing stage or gets unloaded from the system. Neither is possible for the processes in the process subset $\mathcal{DP}$ implied by Definition 7. Thus, all the processes included in $\mathcal{DP}$ cannot proceed to completion, and therefore, the RAS empty set s_0 is unreachable from s. ◇

PROPOSITION 2 *In the RAS-STD, every directed path that starts from an unsafe state, s, and does not involve the loading of a new process in the system, results in a deadlock.*

Proof: Suppose not. Let s_i be an unsafe state on one of the paths emanating from s. Since, by the working hypothesis, s_i is not a deadlock, by taking as $\mathcal{DP}^i$ the entire set of process instances in the system, it follows that there will exist a process instance whose immediate resource requirements are smaller than the current resource slacks, and therefore, it is able to proceed to its next processing stage. Let s_{i+1} denote the resulting state. By invoking the same argument repeatedly, we can keep reducing the total workload in the system by one unit (i.e., one processing stage) at a time. Given the finiteness of the process routes and assuming that no new process instances are loaded into the system, the system is going to be empty after a sufficient number of events. But this contradicts the definition of unsafety and concludes the proof. ◇

It should be noticed, however, that there can be unsafe states that are not deadlocks. As an example, consider state s_{15} in the STD of Figure 2.1: This state, although one step away from deadlock, it does not contain a deadlock itself, since both of the running process instances can advance to their next requested resource R_2. These states will be referred to as *deadlock-free unsafe* states and, as it will be shown in the following, they are the main source for the non-polynomial complexity of the maximally permissive nonblocking supervision in the considered RAS class. Before, however, getting into these complexity considerations, we need to formally characterize the concept of (maximally permissive) nonblocking supervisor for the considered RAS class and the corresponding logical control problem.

2.3   Optimal Nonblocking Supervisory Control

The characterization of state safety and its relationship to the RAS deadlock provided in the previous section, facilitates the formal definition of *deadlock avoidance*, which is the deadlock resolution strategy of interest in this book. In the following discussion, it is assumed that the RAS always undergoes normal operation, i.e., its operation is contained in the reachable subspace S_r.

In the FSA modelling context, a *nonblocking supervisory control policy (SCP)* must restrict the operation of a given RAS by limiting it to its reachable and safe subspace S_{rs}. Practically, we seek to identify an appropriate set of feasible transitions which when removed from the STD – or equivalently, *disabled* by the SCP – render the unsafe subspace $S_{r\bar{s}}$ unreachable from state s_0. At the same time, it must be ensured that every state s in the remaining graph – i.e., *reachable under the control policy* – is still safe – i.e., there exists a directed path *in the remaining graph* leading from state s to s_0. States which are reachable under an enforced SCP and from which progress is inhibited by the policy-imposed constraints and not by the RAS structure, are characterized as *(policy-)induced* or *restricted* deadlocks in the literature (Banaszak and Krogh, 1990).

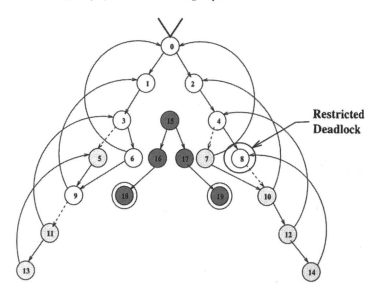

Figure 2.2. Example: An SCP inducing restricted deadlock

An example of an SCP that gives rise to restricted deadlock is presented in Figure 2.2. This hypothetical policy, defined on the RAS-STD of Figure 2.1, *admits*[2] only the states corresponding to the white-colored nodes in the depicted STD. Notice that the policy provides accessibility to state s_8, while it disables the only transition out of it, by not admitting state s_{10}. As a result, whenever the system finds itself in state s_8, it is permanently blocked there by the policy logic itself.

A formal characterization of the above concepts is as follows:

DEFINITION 8 *Consider a RAS* $\Phi =< \mathcal{R}, C, \mathcal{P}, \mathcal{A}, \mathcal{T} >$, *modelled by the FSA* $G =< E, S, f, \Gamma, s_0, S_m >$. *A (logical) supervisory control policy (SCP)* Δ *for it is a function*

$$\Delta : S \to 2^E, \text{ with } \Delta(s) = \{e \in \Gamma(s) : e \text{ is selected by the policy}\}$$

Events $e \in \bigcup_{s \in S} \Delta(s)$ *are called the* policy-enabled *or* admissible *events.*

DEFINITION 9 *Given an SCP* Δ, *let* $s' \overset{\Delta}{\Leftarrow} s$ *denote the fact that state* s' *is reachable from state* s *through an event sequence which comprises policy-enabled events only. Let* $S_r(\Delta) = \{s : s \overset{\Delta}{\Leftarrow} s_0\}$ *and* $S_s(\Delta) = \{s : s_0 \overset{\Delta}{\Leftarrow} s\}$. *Then, policy* Δ *is* non-blocking *or* correct *iff* $S_r(\Delta) \subseteq S_s(\Delta)$.

---

[2]i.e., allows access to

In words, an SCP is correct *iff* the policy-reachable subspace $S_r(\Delta)$ is a *strongly connected subgraph containing the idle and empty state* s_0. Notice that the reachable and safe subspace of the uncontrolled system, S_{rs}, possesses this property; in fact, this is the *maximal* subspace possessing this property. This leads us to the concept of the *maximally permissive* or *optimal* SCP: A correct SCP Δ^* is *optimal iff* the policy restriction on S_{rs} disables *only* those actions that result in unsafe states. Formally,

DEFINITION 10 *A correct SCP* Δ^* *is* optimal *iff*

$$\forall s \in S_{rs}, \ \forall e \in \Gamma(s), \ e \in \Delta^*(s) \Longleftrightarrow f(s,e) \in S_s$$

This characterization of the optimal policy has the following three implications:

1 For a given RAS configuration, the optimal nonblocking SCP Δ^* is *unique*.

2 $S_r(\Delta^*) = S_{rs}$. Establishing the optimal control policy Δ^* is equivalent to removing from the reachability graph those transition arcs that belong to the STD *cut* $[S_{rs}, S_{r\bar{s}}]$. For example, in the STD of Figure 2.1, the optimal control policy Δ^* consists of removing the arcs that emanate from lightly shaded solid nodes and cross the twisted dashed line.

3 From a computational standpoint, resolving the admissibility of a feasible transition by the optimal SCP Δ^* requires the assessment of the safety of the resulting state; i.e., given a RAS state $s \in S_r$ and a tentative feasible transition corresponding to some labelling event $e \in \Gamma(s)$, the controller implementing the optimal SCP must (i) simulate the execution of transition $f(s,e)$ on the underlying RAS and obtain the resultant state s', (ii) assess the safety of s', and (iii) admit the transition *iff* $s' \in S_s$.

The mechanism described in Item 3 above for implementing the optimal SCP belongs to a broader class of SCP's, known as *one-step-lookahead* policies. A formal characterization of this concept is as follows:

DEFINITION 11 *An SCP* Δ *for a given RAS* $\Phi = < \mathcal{R}, C, \mathcal{P}, \mathcal{A}, \mathcal{T} >$ *is characterized as* one-step-lookahead, *iff there exists a property* $\mathcal{H}$ *defined on the RAS state space* S, *such that*

$$\forall s \in S, \ \forall e \in \Gamma(s), \ e \in \Delta(s) \Longleftrightarrow f(s,e) \in \mathcal{H}(s)$$

The next section establishes that the RAS *safety* decision problem – i.e., "Given s, is $s \in S_s$?" – is NP-complete (Garey and Johnson, 1979) in the considered RAS class, and therefore, the implementation of the optimal SCP will be a computationally intractable problem for most practical applications.

3. The computational complexity of the Optimal Nonblocking SCP

As it was established in the previous section, the computational complexity of the optimal SCP for the considered RAS class is determined by the complexity of the *safety* problem, i.e., given a RAS state s, does there exist a resource allocation sequence that can advance all running processes to completion? In this section we show that, in the general case, this problem belongs to the notorious class of NP-complete problems (Garey and Johnson, 1979). Some results on the computational complexity of the RAS safety problem appeared initially in (Araki et al., 1977; Gold, 1978). The result reported next comes from (Lawley and Reveliotis, 2001), and it can be claimed to be the most general with respect to the RAS classes considered in this book, since it establishes the NP-completeness of safety even for the case of LIN-SU-RAS; we remind the reader that the RAS behavior generated by this class is subsumed by any other class of the RAS taxonomy introduced in Chapter 1.

THEOREM 2.1 *The problem of RAS safety is NP-complete even for RAS in which each process constitutes a linear sequence of processing stages with each stage requesting a single unit from a single resource type for its execution.*

Proof: First of all, notice that safety belongs to NP, since we can verify the validity of any candidate termination sequence, through simulation, in time $O(|\mathcal{R}| \cdot \max_j\{|\mathcal{S}_j|\} \cdot \sum_i C_i)$.

To prove NP-completeness, we provide a polynomial reduction of the 3-SAT problem to the problem of RAS safety, where the latter is restricted in the RAS domain defined in the theorem. A formal statement of the 3-SAT problem is as follows (Garey and Johnson, 1979):

DEFINITION 12 *Let* $\chi = \{X_1, \bar{X}_1, X_2, \bar{X}_2, \ldots, X_\mu, \bar{X}_\mu\}$ *be a set of literals and* $\Lambda = \Lambda_1 \wedge \Lambda_2 \wedge \ldots \wedge \Lambda_\nu$ *be a conjunction of clauses of the form* $\Lambda_q = Y_{q1} \vee Y_{q2} \vee Y_{q3}$, *where* $Y_{qk} \in \chi$, *i.e., each* Y_{qk} *is one of the literals belonging to* χ. *Let* $K \subseteq \chi$ *be such that*

CONDITION 4 $\forall i = 1, \ldots, \mu$, *either* $X_i \in K$ *or* $\bar{X}_i \in K$, *but not both.*

Then, the satisfiability question is as follows: Given χ *and* Λ, *does there exist* $K \subseteq \chi$ *satisfying Condition 4 such that* $\forall q = 1, \ldots, \nu$, $K \cap \Lambda_j \neq \emptyset$?

An instance of the 3-SAT problem, defined by $< \chi, \Lambda >$, can be reduced polynomially to the considered safety problem as follows:

For each clause $\Lambda_q \in \Lambda$, define seven processes Π_{q1} to Π_{q7}, with the following resource sequences

$$\Pi_{q1} = < \Lambda_{q1}, \Lambda_{q4}, \Lambda_{q5}, \Lambda_{q6}, \Lambda_{q8}, Y_{q1} >, \quad \Pi_{q5} = < \Lambda_{q6}, \Lambda_{q8}, \Lambda_{q1}, \Lambda_{q2}, \Lambda_{q3} >$$

$$\Pi_{q2} = <\Lambda_{q2}, \Lambda_{q4}, \Lambda_{q5}, \Lambda_{q6}, \Lambda_{q8}, Y_{q2}>, \qquad \Pi_{q6} = <\Lambda_{q8}, \Lambda_{q1}, \Lambda_{q2}, \Lambda_{q3}>$$
$$\Pi_{q3} = <\Lambda_{q3}, \Lambda_{q4}, \Lambda_{q5}, \Lambda_{q6}, \Lambda_{q8}, Y_{q3}>, \qquad \Pi_{q7} = <\Psi_q, \Lambda_{q8}, \Lambda_{q6}, \Lambda_{q7}>$$
$$\Pi_{q4} = <\Lambda_{q7}, \Lambda_{q6}, \Lambda_{q8}, \Lambda_{q1}, \Lambda_{q2}, \Lambda_{q3}>$$

Note that for each clause, Λ_q, we define twelve resources $\{\Lambda_{q1} \ldots, \Lambda_{q8}, Y_{q1}, Y_{q2}, Y_{q3}, \Psi_q\}$. Furthermore, in agreement with the definition of the 3-SAT problem, each resource Y_{qk}, $k = 1, 2, 3$, represents some X or $\bar{X}$ in χ. Next, for each pair of 3-SAT literals, X_i and $\bar{X}_i$, define a pair of processes, Π_{i1} and Π_{i2}, with the following resource sequences:

$$\Pi_{i1} = <X_i, B_i, D_1, D_2> , \quad \Pi_{i2} = <\bar{X}_i, B_i, D_1, D_2>$$

Note that for each i, we define resources $\{X_i, \bar{X}_i, B_i\}$. We also define two "global" resources $\{D_1, D_2\}$. Next, define two processes, Π_{01} and Π_{02}, with resource sequences:

$$\Pi_{01} = <D_1, \Psi_1, \Psi_2, \ldots, \Psi_{\nu-1}, \Psi_\nu>$$

$$\Pi_{02} = <D_2, D_1, \Psi_\nu, \Psi_{\nu-1}, \ldots, \Psi_2, \Psi_1>$$

Finally, advance every process to its first stage in the corresponding sequence; the resulting RAS state is depicted in Figure 2.3. Note that the number of processes defined is $2\mu + 7\nu + 2$, the number of resources is $3\mu + 9\nu + 2$, and μ and ν were respectively defined in Definition 12 as the cardinality of set χ and the number of clauses. Therefore, the presented reduction is polynomial.

Now, suppose that Λ is satisfiable. Then there exists K such that $K \cap \Lambda_q \neq \emptyset$, $\forall q = 1, \ldots, \nu$. For each $i = 1, \ldots, \mu$, if $X_i \in K$, advance Π_{i1} to B_i; otherwise, advance Π_{i2} to B_i. Figure 2.4 gives the resulting state. Notice that, for each $q = 1, \ldots, \nu$, at least one resource in the set $\{Y_{q1}, Y_{q2}, Y_{q3}\}$ is free. Next we show that for each $q = 1, \ldots, \nu$, the processes $\Pi_{q1}, \Pi_{q2}, \Pi_{q3}, \Pi_{q4}, \Pi_{q5}, \Pi_{q6}$ and Π_{q7} can be advanced to a point where Ψ_q is released without incurring unsafeness. We consider three cases:

Case 1: $Y_{q1} \in K \cap \Lambda_q$. If Y_{q1} is free, advance processes $\Pi_{q1}, \Pi_{q2}, \Pi_{q3}, \Pi_{q4}, \Pi_{q5}, \Pi_{q6}$ and Π_{q7} as follows: Π_{q1} to Λ_{q4} to Λ_{q5}; Π_{q2} to Λ_{q4}; Π_{q6} to Λ_{q1} to Λ_{q2}; Π_{q5} to Λ_{q8} to Λ_{q1}; Π_{q1} to Λ_{q6} to Λ_{q8} to Y_{q1} and out of the system; Π_{q2} to Λ_{q5}; Π_{q3} to Λ_{q4}; Π_{q6} to Λ_{q3} and out of the system; Π_{q5} to Λ_{q2} to Λ_{q3} and out of the system; Π_{q4} to Λ_{q6} to Λ_{q8} to Λ_{q1} to Λ_{q2} to Λ_{q3} and out of the system; Π_{q7} to Λ_{q8} to Λ_{q6} to Λ_{q7} and out of the system. Figure 2.5(a) shows the resulting state. Note that every Π_q. has finished, except for Π_{q2} and Π_{q3}, and that resource Ψ_q is free.

Case 2: $Y_{q2} \in K \cap \Lambda_q$. If Y_{q2} is free, advance processes $\Pi_{q1}, \Pi_{q2}, \Pi_{q3}, \Pi_{q4}, \Pi_{q5}, \Pi_{q6}$ and Π_{q7} as follows: Π_{q2} to Λ_{q4} to Λ_{q5}; Π_{q1} to Λ_{q4}; Π_{q6} to Λ_{q1} to Λ_{q2}; Π_{q5} to Λ_{q8} to Λ_{q1}; Π_{q2} to Λ_{q6} to Λ_{q8} to Y_{q2} and out of the system; Π_{q1} to Λ_{q5}; Π_{q3} to Λ_{q4}; Π_{q6} to Λ_{q3} and out of the system; Π_{q5} to Λ_{q2} to Λ_{q3} and out of the

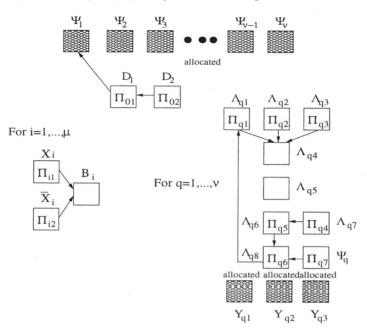

Figure 2.3. The RAS state induced by the 3-SAT problem

system; Π_{q4} to Λ_{q6} to Λ_{q8} to Λ_{q1} to Λ_{q2} to Λ_{q3} and out of the system; Π_{q7} to Λ_{q8} to Λ_{q6} to Λ_{q7} and out of the system. Figure 2.5(b) shows the resulting state. Note that every Π_{q}. has finished, except for Π_{q1} and Π_{q3}, and that resource Ψ_q is free.

Case 3: $Y_{q3} \in K \cap \Lambda_q$. If Y_{q3} is free, advance processes $\Pi_{q1}, \Pi_{q2}, \Pi_{q3}, \Pi_{q4}$, Π_{q5}, Π_{q6} and Π_{q7} as follows: Π_{q3} to Λ_{q4} to Λ_{q5}; Π_{q1} to Λ_{q4}; Π_{q6} to Λ_{q1}; Π_{q5} to Λ_{q8}; Π_{q3} to Λ_{q6}; Π_{q1} to Λ_{q5}; Π_{q2} to Λ_{q4}; Π_{q6} to Λ_{q2} to Λ_{q3} and out of the system; Π_{q5} to Λ_{q1} to Λ_{q2} to Λ_{q3} and out of the system; Π_{q3} to Λ_{q8} to Y_{q3} and out of the system; Π_{q4} to Λ_{q6} to Λ_{q8} to Λ_{q1} to Λ_{q2} to Λ_{q3} and out of the system; Π_{q7} to Λ_{q8} to Λ_{q6} to Λ_{q7} and out of the system. Figure 2.5(c) shows the resulting state. Note that every Π_{q}. has finished, except for Π_{q1} and Π_{q2}, and that resource Ψ_q is free.

Therefore, for $q = 1, \ldots, \nu$, it is possible to complete all processes in the set $\{\Pi_{q4}, \Pi_{q5}, \Pi_{q6}, \Pi_{q7}\}$, thus releasing resource Ψ_q. Furthermore, it is possible to complete at least one of Π_{q1}, Π_{q2} or Π_{q3}, with the remaining processes being in one of the states in Figure 2.5. Since Ψ_q is now free, for $q = 1, \ldots, \nu$, advance Π_{01} to $\Psi_1, \Psi_2, \ldots, \Psi_\nu$ and out of the system. Next, advance Π_{02} to $D_1, \Psi_\nu, \Psi_{\nu-1}, \ldots, \Psi_1$ and out of the system. Now, since D_1 and D_2 are free, each Π_{i1} and Π_{i2} can be advanced to D_1, D_2 and out of the system,

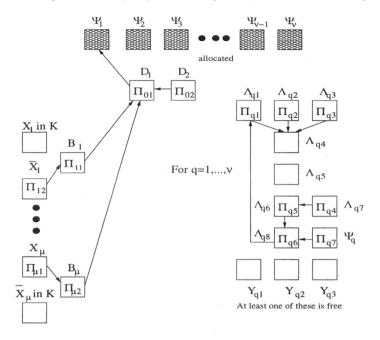

Figure 2.4.   The RAS state after the release of all resources in K

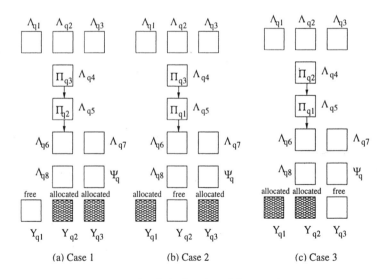

Figure 2.5.   A safe deallocation of Ψ_q, $q = 1, \dots, \nu$

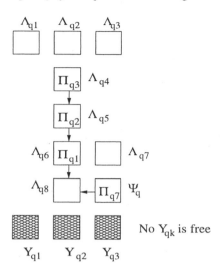

Figure 2.6. The case where $\Lambda_q \cap K = \emptyset$

for $i = 1, \ldots, \mu$, thus releasing all resources X_i and $\bar{X}_i$. Therefore, every resource in the set $\{Y_{q1}, Y_{q2}, Y_{q3}\}$ is free for $q = 1, \ldots, \nu$. Hence, for every $q = 1, \ldots, \nu$, advance all remaining Π_{qk} on Λ_{q5} to $\Lambda_{q6}, \Lambda_{q8}, Y_{qk}$ and out of the system. Then, advance all remaining Π_{qk} on Λ_{q4} to $\Lambda_{q5}, \Lambda_{q6}, \Lambda_{q8}, Y_{qk}$ and out of the system. Since all processes are completed and all resources are released, the state of Figure 2.3 is safe.

Now suppose that Λ is not satisfiable. We shall show that the state of Figure 2.3 is not safe. Since Λ is not satisfiable, for every $K \subseteq \chi$ satisfying Condition 4, there exists Λ_q such that $K \cap \Lambda_q = \emptyset$. Without loss of generality, consider a given K and $i \in \{1, \ldots, \mu\}$. If $X_i \in K$, advance Π_{i1} to B_i; otherwise, advance Π_{i2} to B_i. Next we show that, starting from the state of Figure 2.4, if Π_{q7} releases Ψ_q before a Y_{qk} is available, deadlock results. Note that if Π_{q7} advances to Λ_{q8} (and releases Ψ_q) before Π_{q4} advances to Λ_{q1}, deadlock involving Π_{q7} and Π_{q4} is inevitable. Therefore, we must advance processes so that Π_{q4} can gain Λ_{q1}. Starting from the state given in Figure 2.4 and without loss of generality, advance the processes as follows: Π_{q1} to Λ_{q4} to Λ_{q5}; Π_{q2} to Λ_{q4}; Π_{q6} to Λ_{q1} to Λ_{q2}; Π_{q5} to Λ_{q8} to Λ_{q1}; Π_{q4} to Λ_{q6} to Λ_{q8}; Π_{q1} to Λ_{q6}; Π_{q2} to Λ_{q5}; Π_{q3} to Λ_{q4}; Π_{q6} to Λ_{q3} and out of the system; Π_{q5} to Λ_{q2} to Λ_{q3} and out of the system; Π_{q4} to Λ_{q1} to Λ_{q2} to Λ_{q3} and out of the system. These advancements yield the state of Figure 2.6. Note that Π_{q1} cannot proceed beyond Λ_{q8} since none of the Y_{qk} are available. At this point, if Π_{q7} advances to Λ_{q8} and releases Ψ_q, it deadlocks with Π_{q1}. Therefore, if $K \cap \Lambda_q = \emptyset$, Ψ_q must not be released, otherwise, Π_{q7} becomes involved in deadlock.

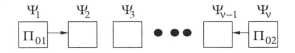

Figure 2.7. The unsafe state involving Π_{01} and Π_{02}

Therefore, in the state of Figure 2.4, there exists at least one set of processes $\{\Pi_{q1}, \Pi_{q2}, \Pi_{q3}, \Pi_{q7}\}$ that cannot be completed until additional Y_{qk}'s are released. Furthermore, the corresponding Ψ_q will not be released. To release additional Y_{qk}'s, we must free some B_i resources by advancing P_{01}. If $K \cap \Lambda_1 = \emptyset$, Ψ_1 is not available. Π_{01} cannot advance, no other Y_{qk}'s can be released, at least one set of processes of the form $\{\Pi_{q1}, \Pi_{q2}, \Pi_{q3}, \Pi_{q7}\}$ cannot be completed, and the system is unsafe. If Ψ_1 is available, advance Π_{01} to Ψ_1. We must now advance Π_{02} to D_1, for if we advance any Π_i from B_i to D_1, it deadlocks with Π_{02}. Our only next choice is to advance Π_{02} to Ψ_ν, thus freeing D_1. If $K \cap \Lambda_\nu = \emptyset$, then Ψ_ν is not available, P_{02} cannot be advanced, no other Y_{qk}'s can be released, at least one set of processes of the form $\{\Pi_{q1}, \Pi_{q2}, \Pi_{q3}, \Pi_{q7}\}$ cannot be completed, and the system is unsafe. If Ψ_ν is available, advance Π_{02} to Ψ_ν. D_1 and D_2 are now free, so all Π_i processes can be completed, one at a time, and all resources X_i and $\bar{X}_i$ can be released. Therefore, for $q = 1, \ldots, \nu$, all process sets $\{\Pi_{q1}, \Pi_{q2}, \Pi_{q3}, \Pi_{q4}, \Pi_{q5}, \Pi_{q6}, \Pi_{q7}\}$ can be completed, and all resources Ψ_q can be released. Only Π_{01} and Π_{02} remain to be completed (see Figure 2.7). Unfortunately, Π_{01} and Π_{02} are headed for an inevitable deadlock. To see this, note that Π_{01} holds Ψ_1 and requires the resource sequence $< \Psi_2, \Psi_3, \ldots, \Psi_{\nu-1}, \Psi_\nu >$, while Π_{02} holds Ψ_ν and requires the resource sequence $< \Psi_{\nu-1}, \Psi_{\nu-2}, \ldots, \Psi_2, \Psi_1 >$. Therefore, it is impossible to complete these two processes and the system is unsafe. $\diamond$

The next theorem establishes that, while RAS safety is an NP-complete problem, the RAS deadlock itself is polynomially recognizable. Therefore, it can be concluded that *the super-polynomial complexity of the RAS safety problem is due to the difficulty of recognizing deadlock-free unsafe states.*

THEOREM 2.2 *In the class of Disjunctive / Conjunctive RAS, the problem of determining whether a given RAS state is a deadlock or not, is of polynomial complexity with respect to the underlying RAS size.*

Proof: An algorithm for resolving this problem is presented in Figure 2.8. This algorithm scans repetitively the non-zero components of the state vector s, setting each of them to 0 upon the identification of a resource allocation request that (i) enables the transition of the corresponding process instances to one of their successor processing stages, and (ii) it can be satisfied from the resource slacks corresponding to the running value of s. If the algorithm succeeds to set

Deadlock Detection Algorithm for DIS-CON-RAS
Input: DIS-CON-RAS $\Phi = < \mathcal{R}, C, \mathcal{P}, \mathcal{A}, \mathcal{T} >$ and a state s
Output: Boolean variable DEADLOCK

```
begin
  /* Initialize */
  s' := s
  DEADLOCK := FALSE;

  /* processing step */
  while (s' ≠ 0 and not(DEADLOCK)) do
  begin
    DEADLOCK := TRUE;
    for (q:=1 to dim(s')) do
    begin
      if (s'(q) > 0)
      begin
        succ:=< ordered list of indices to successor stages for q >;
        while (DEADLOCK and succ ≠ NULL) do
        begin
          p:= head(succ); succ:=pop(succ,p);
          i:=1; DEL:=TRUE;
          while (i ≤ dim(R) and DEL) do
          begin
            DEL := (δᵢ(s') ≥ A₍ₚ₎(i) − A₍q₎(i));
            i:=i+1;
          endwhile
          if (DEL)
          begin
            s'(q) := 0;
            DEADLOCK := FALSE;
          end
        endwhile
      end
    endfor
  endwhile

end
```

Figure 2.8. A polynomial deadlock detection algorithm for the Disjunctive / Conjunctive RAS

all the state components to 0, the considered state s is deadlock-free; otherwise, s is declared to be a deadlock.

Next we show that this algorithm is of polynomial complexity. The algorithm must "delete" – i.e., set equal to 0 – at least one state component in order to proceed to its next scan of the state vector. Therefore, the maximum number of "deletion" tests performed by the algorithm, throughout its entire operation, is $O((\dim(s))^2) = O((\sum_{j=1}^{n} |S_j|)^2)$. The number of successor stages that must be examined at each of those "deletion" tests is of order $O(\max_{j=1}^{n} |S_j|)$, while the computational cost for assessing the feasibility of the transition from a given stage q to a given stage p is $O(|\mathcal{R}|)$. Therefore, the overall complexity of the algorithm is $O(|\mathcal{R}| \max_{j=1}^{n} |S_j| (\sum_{j=1}^{n} |S_j|)^2)$. $\diamond$

4.   Practical approaches to the RAS logical control problem

As it has been already pointed out, from a practical standpoint, the result of Theorem 2.1 is a negative result; it establishes that, in the general case, the optimal SCP will be computationally intractable in the operational context of the considered RAS classes. The typical reaction in the face of such a result is either (i) to identify *special structure* for the problem in which the target concept – in our case, the optimal SCP – is polynomially computable, or (ii) to try to generate "good" *approximations* of it that can be obtained with polynomial computational cost. These are the tracks that have been followed by the research community in this case, as well. Special RAS structure that enables the implementation of the optimal SCP in polynomial complexity w.r.t. the RAS size, is discussed in Chapter 3. On the other hand, the notion of a "good" polynomial approximation to the optimal SCP has given rise to the concept of *Polynomial Kernel (PK-)* SCP's. Simply stated, the basic idea behind PK-SCP's is that, since the target set S_s is not polynomially recognizable, the system operation should be confined to a subset of these states which is polynomially computable. This state subset is perceived as an easily identifiable – i.e., polynomially computable – *kernel* among the set of reachable and safe states, and gives the methodology its name.

From an implementational viewpoint, the development of PK-SCP's requires the identification of a property $\mathcal{H}(s)$, $s \in S$, such that: (i)$\mathcal{H}(s_0) = $ TRUE, (ii) $\mathcal{H}(s) \Rightarrow $ safe$(s), \forall s \in S_r$, and (iii) $\mathcal{H}()$ is *polynomially* testable on the system states. Then, by allowing only transitions to states satisfying $\mathcal{H}$, through one-step lookahead, it can be ensured that the visited states will be safe. An additional important requirement is that (iv) the resulting SCP is *correct*, i.e., the policy-reachable subspace must be strongly connected (cf. Section 2.3, Definition 9). However, this characterization of policy correctness is based on a global view of the system operation, and given the typically large size of the system state space, it is not easily verifiable. A more operable criterion for testing the correctness of Polynomial-Kernel policies is provided by the following theorem:

THEOREM 2.3 *A Polynomial-Kernel SCP is* correct *iff for every state admitted by the policy, there exists a policy-admissible transition, and this transition does not correspond to a loading event.*

The validity of this theorem results from the observation that at any point in time, the system workload – in terms of processing steps – is finite, and every transition described in the theorem reduces this workload by one unit. Hence, eventually the total workload will be driven to zero, which implies that the system has returned to its marked state s_0. A more formal statement and proof of this theorem, by means of an algebraic characterization of state safety, can be found in (Reveliotis and Ferreira, 1996). It should be noticed that establishing the policy correctness by means of Theorem 2.3, resolves concurrently the validity of condition $\mathcal{H}$ as a Polynomial-Kernel identifier for state safety, and the restricted deadlock-free operation of the controlled system.

A last concern is that the developed PK-SCP's are *efficient*, i.e., they provide considerable flexibility for the system operation, by admitting an extensive part of S_{rs}. The design of efficient PK-SCP's that are appropriate for the various classes of the RAS taxonomy defined in Section 2.2 of Chapter 1, is undertaken in Chapters 4 and 5.

5. Some extensions of the RAS logical control problem

In this section we extend the nonblocking SC problem for DIS-CON-RAS in order to account for additional elements in the underlying system behavior. More specifically, the first of these extensions considers the problem of establishing nonblocking supervision when the behavior of the underlying RAS is *uncontrollable* with respect to certain resource allocations. The second extension deals with the problem of *(re-)configuration management*, i.e., the preservation of nonblocking operation in the face of changes in the system structure, e.g., due to resource failures. Since both of these extended versions subsume the basic problem definition considered in the earlier sections of this chapter, it can be concluded that the development of maximally permissive SC policies for them will be computationally intractable. Indeed, the problem of synthesizing effective and computationally efficient SC policies for these extended RAS classes is a challenging problem currently under active investigation. Here, we provide detailed characterizations of the aforementioned problems by means of the FSA modelling framework; some additional available results on them are reported in later chapters.

5.1 Nonblocking supervision under uncontrollable resource allocations

In the entire development of this chapter up to this point, we have implicitly assumed that (i) it is upon the exclusive discretion of the RAS controller to disable and/or postpone the execution of the various feasible events during the system operation, and that (ii) process instances with flexible routing can be forced, if necessary, to a certain routing option. Here we seek to relax these two assumptions. In order to motivate this need, we notice that, in certain cases, the system might not have the necessary hardware that would facilitate the disabling of certain allocation events. In other cases, the inability of the controller to postpone the execution of certain resource allocations, once they are physically possible, can result from the time-criticality of the involved operations; as a more concrete example for this case, taken from the manufacturing domain, consider a forging operation the must follow the preheating of the part at a certain temperature. Both of these cases refute the first assumption mentioned above, regarding the controllability of the execution of the various feasible events, and they will be collectively characterized as *Type-1 uncontrollablity*. The second assumption, regarding the controllability of the process routing, will be naturally violated by systems with non-perfect yield: in these contexts, the eventual routing of each process instance will be contingent upon the outcome of its current processing stage, and it will not be (exclusively) determined by the controller. We shall refer to this type of uncontrollability, resulting from uncontrollable routing, as *Type-2 uncontrollability*.

Our intention is to develop an SCP that will maintain nonblocking operation for the underlying RAS in spite of the occurrence of Type-1 and/or Type-2 uncontrollable events. We start by providing the formal specifications for such a supervisor in the FSA-modelling context. To facilitate the subsequent discussion, let E_u^1 and E_u^2 respectively denote the Type-1 and Type-2 uncontrollable events of the considered RAS; $E \backslash (E_u^1 \cup E_u^2) \equiv E_c$ will be the set of *controllable* events.

DEFINITION 13 *An SCP* Δ *is* acceptable *w.r.t.* E_u^1 *iff* $\forall s \in S$, $(e \in \Gamma(s)) \wedge (e \in E_u^1) \Longrightarrow e \in \Delta(s)$.

DEFINITION 14 *An SCP* Δ *is* acceptable *w.r.t.* E_u^2 *iff* $\forall s \in S$, $(e \in \Gamma(s)) \wedge (e \in E_u^2) \Longrightarrow \exists$ *a state sequence* $s \xrightarrow{\Delta(s)} s_1 \xrightarrow{\Delta(s_1)} s_2 \xrightarrow{\Delta(s_2)} \ldots \xrightarrow{\Delta(s_{n-1})} s_n$ *s.t.*

1 $e \in \Delta(s_n)$;

2 *if* e *corresponds to the advancement of stage* Ξ_{jk}, *then,* $\forall \nu = 1, \ldots, n$, $s_\nu(q(j,k)) \geq s(q(j,k))$.

In words, an acceptable SCP Δ cannot delay the occurrence of Type-1 uncontrollable events, and it cannot prevent the occurrence of Type-2 uncontrollable events, even though it might postpone the latter until some other events have occurred. Under the reasonable assumptions that (i) all loading events are controllable, and (ii) resource allocation vectors are well-defined w.r.t. the resource

Input: FSA $G = < E_c \cup E_u^1 \cup E_u^2, S, f, \Gamma, s_0, \{s_0\} >$ representing a DIS-CON-RAS
Output: Maximally permissive acceptable nonblocking SCP Δ^* and the policy-admissible space $S_r(\Delta^*)$

1 Generate the reachability space S_r for the considered RAS.

2 Extract the reachable and safe subspace S_{rs}. One effective way to execute this operation is by reversing the arcs in the STD corresponding to S_r, and marking all states that can be reached from s_0 in the modified STD; the marked nodes correspond to S_{rs}.

3 Consider the policy Δ that restricts the system operation in S_{rs}, and compute the set $S_D \subseteq S_{rs}$ for which Δ fails to satisfy the acceptance condition of Definition 13.

4 If $S_D = \emptyset$, proceed to Step 5; otherwise, set $S_r := S_{rs} \backslash S_D$, and go to Step 2.

5 Consider the policy Δ that restricts the system operation in S_{rs}, and compute the set $S_D \subseteq S_{rs}$ for which Δ fails to satisfy the acceptance condition of Definition 14.

6 If $S_D = \emptyset$, proceed to Step 7; otherwise, set $S_r := S_{rs} \backslash S_D$, and go to Step 2.

7 Return $\Delta^* := \Delta$ and $S_r(\Delta^*) := S_{rs}$.

Figure 2.9. An algorithm computing the maximally permissive acceptable nonblocking SCP for DIS-CON-RAS with Type-1 and/or Type-2 uncontrollability

capacities – i.e., $A_{jk}(i) \leq C_i$, $\forall i, j, k$ – such a policy will always exist: this is the policy that allows only one process instance in the system. Next, we characterize also the *maximally permissive* nonblocking supervisor Δ^* for this extended RAS class. In particular, Figure 2.9 presents an algorithm that, when executed on any given RAS belonging to the considered sub-class, will return the maximally permissive nonblocking acceptable SCP Δ^*.

The algorithm of Figure 2.9 starts with the reachable and safe space S_{rs}, that characterizes the system behavior under the maximally permissive SCP, when the system is totally controllable, and subsequently, it removes all states that fail to satisfy the acceptance condition of Definition 13 w.r.t. the policy Δ induced by S_{rs}. These are states from which the system can transition uncontrollably to its unsafe region, and therefore, they must be rendered inaccessible by any acceptable policy. However, the removal of these states from S_{rs} will lead to a reduced space S_r that might not be strongly connected and, as a result, the policy Δ induced by it, will not be correct. Hence, the trimming operation of Step 2 must be repeated on the reduced S_r. The resulting S_{rs} might still contain states from which the system can transition uncontrollably to $S \backslash S_{rs}$, and therefore, the loop of Steps 2 and 3 is repeated until the identification of a subspace S_{rs} that is closed under Type-1 uncontrollability. At this point the satisfaction of the requirement of Definition 14 is taken on. If it happens that the subspace S_{rs} developed in the last execution of Step 2 satisfies also the requirement of Definition 14, then there is nothing more to be done; if, however, there are policy-admissible states violating this new condition, then they must be eliminated, and the algorithm returns to Step 2, in order to generate a more restrictive policy that satisfies the correctness and acceptance requirements of Definitions 9, 13 and 14. A more formal proof for the correctness of this algorithm can be obtained through the linguistic frameworks of (Cassandras and Lafortune, 1999; Kumar and Garg, 1995).

The computation involved in the above algorithm also establishes that Δ^* is uniquely defined. However, as it was indicated in the opening discussion of this section, the result of Theorem 2.1 implies that the implementation of Δ^* is an NP-hard problem (Garey and Johnson, 1979). A methodology that generates polynomial, correct and acceptable, sub-optimal SCP's for DIS-CON-RAS with Type-1 and/or Type-2 uncontrollability, is presented in Section 5 of Chapter 5.

5.2   Maintaining nonblocking operation under resource failures

All the above analysis has presumed a stable system configuration; in particular, it has been assumed that the resource capacities and the set of job types supported by the system does not change over time. In this section we consider how to effectively accommodate in the applied control logic any changes in these two elements of the system configuration. From a conceptual standpoint, there are two main approaches that can be adopted with respect to this issue:

The reactive approach   As suggested by its name, this approach simply reacts to the various contingencies taking place in the system, by developing a new control policy for the emerging RAS configuration. Hence, in the context of this approach, it is important that (i) the applied SCP's can be algorithmically developed for any given RAS configuration Φ, and furthermore, (ii) the complexity of the algorithms involved in the policy development is fairly low. An additional consideration is to design the new policy in a way that it can accommodate the running RAS state with minimal disruption, i.e., the number – or, more generally, a cost measure determined by the set – of active processes that must be rearranged or unloaded from the system, in order to obtain a RAS state admissible by the new policy, must be minimized. In general, this approach is more appropriate for systems in which the aforementioned disruptions, and the associated need for policy reconfiguration, are rather rare.

The proactive approach Under this approach, there is one single policy that controls the system operation under all emerging configurations. The applied control logic anticipates potential disruptions in the system operation, and it seeks to ensure that during the occurrence of any such disruption, the system operation will degrade in a *"graceful"* manner. In the particular case of a resource failure, the applied policy logic seeks to ensure that the system will still be able to maintain the execution of all those process types that are not requesting the failing resource(s) for their execution. This is to be achieved without any external intervention, but simply by exploiting those resources whose operation is necessarily stalled because of the experienced failure(s), as a buffer that will accommodate all the process instances blocked in the system because of their need to use the failing resources. A major challenge for this approach is to distribute these blocked jobs in the aforementioned buffering resources in a way that the resulting RAS state will also be policy-admissible in the original RAS configuration, that will be re-established by the repair of the failing resources. The proactive approach seems to be more appropriate for systems experiencing frequent outages, and therefore, the cost / effort of reconfiguring the RAS state and the applied SCP would be too high. On the other hand, by accounting for the "worst-case" scenario at any point in time, SCP's following the proactive approach are quite restrictive / conservative in terms of their permissiveness. From a methodological standpoint, the proactive approach is an application of *robust control* on the RAS supervisory control problem.

Some currently available results with respect to both of these approaches are discussed in Chapter 4.

6. Historical and bibliographical notes

As it was mentioned in the bibliographical discussion of Chapter 1, the problem of deadlock avoidance in sequential resource allocation systems has received particular attention in the last decade, primarily because of its emergence as an important problem in the operation of flexibly automated production systems. Some seminal works that introduced the problem in the academic community are those appearing in (Viswanadham et al., 1990; Banaszak and Krogh, 1990; Wysk et al., 1991). All these papers deal with the problem of deadlock avoidance as it arises in LIN-SU-RAS, since their primary motivation is the effective allocation of the buffering capacity in contemporary flexibly automated manufacturing systems. From a methodological standpoint, the work of (Viswanadham et al., 1990) establishes the ability of the FSA modelling framework to provide a systematic characterization of the deadlock avoidance problem. The authors recognize the very high complexity of the problem – although they do not provide any formal results for it – and suggest to deal with this complexity through partial lookahead; their approach tends to reduce the occurrence of deadlock but it does not eliminate it completely. The first well-known provably correct deadlock avoidance policy of polynomial complexity for LIN-SU-RAS was presented in (Banaszak and Krogh, 1990);[3] we shall discuss this policy in detail in Chapter 4. The work of (Wysk et al., 1991) investigates more extensively the structure of the circular dependencies characterizing the deadlock in LIN-SU-RAS and seeks to develop a formal representation and an algorithmic approach for their detection; however, the proposed approach presents non-polynomial computational complexity w.r.t. the RAS size, and therefore, it is not scalable.

The appearance of the aforementioned works was followed by a large number of publications seeking to provide a more profound understanding of the deadlock problem and its complexity, as it appears primarily in the SU-RAS context. Some indicative examples of those works can be found in (Wysk et al., 1994; Hsieh and Chang, 1994; Ezpeleta et al., 1995; Barkaoui and Ben Abdallah, 1995; Reveliotis and Ferreira, 1996; Fanti et al., 1997; Lawley et al., 1997b; Lawley et al., 1997c; Lawley et al., 1998a; Reveliotis et al., 1997; Lawley et al., 1998b; Lawley, 1999; Park and Reveliotis, 2000; Lawley and Reveliotis, 2001). More recently, the research community has effectively addressed the deadlock problem arising in more complex RAS structures; indicative results in this direction are the works presented in (Barkaoui et al., 1997; Tricas et al., 1998; Tricas et al., 1999; Park and Reveliotis, 2001b; Park, 2000; Park and Reveliotis, 2002a; Jeng and Xie, 2001; Lawley and Sulistyono, 2002; Jeng et al., 2002; Reveliotis, 2003a). We shall return to many of these works in the subsequent chapters of this book.

Regarding the material developed in this chapter, the characterization of the DIS-CON-RAS behavior and of the corresponding SC problem through the FSA modelling framework, presented in Sections 1 and 2, is based on (Reveliotis, 1996; Reveliotis and Ferreira, 1996). The complexity result of Theorem 2.1 was published in (Lawley and Reveliotis, 2001). To the best of our knowledge, the algorithm employed in the proof of Theorem 2.2 has not appeared anywhere else, although similar ideas have appeared in other works, e.g., (Peterson, 1981; Reveliotis et al., 1997). The idea of PK-SCP's was originated in (Banaszak and Krogh, 1990) and it was further formalized in (Reveliotis, 1996; Reveliotis and Ferreira, 1996). The concepts of RAS uncontrollability discussed in Section 5.1 have originally appeared in (Park, 2000; Park and Rev-

---

[3]It must be mentioned, however, that many of the ideas underlying the developments of (Banaszak and Krogh, 1990), originally appeared in (Banaszak and Roszkowska, 1988).

eliotis, 2002a). However, the detailed statement of the acceptable nonblocking SC problem in the FSA-based RAS modelling framework, and the algorithm of Figure 2.9 for the computation of the maximally permissive acceptable nonblocking supervisor, is new material. The concepts and ideas regarding the reactive and proactive approaches to RAS re-configuration management are coming respectively from (Reveliotis, 1999) and (Lawley and Sulistyono, 2002).

From a more general standpoint, the FSA-based formulation of the RAS nonblocking SC problem, provided in Sections 1 and 2, is a straightforward implementation of *Ramadge & Wonham's* SC framework (Ramadge and Wonham, 1989) in the RAS operational context. In particular, the work of (Li and Wonham, 1988) has addressed the more abstract problem of developing maximally permissive nonblocking acceptable supervisors for FSA's with uncontrollably enabled events (corresponding to Type-1 uncontrollability in the modelling framework of Section 5.1). The reader is referred to (Cassandras and Lafortune, 1999) for a systematic introduction to Ramadge & Wonham's SC theory.

Chapter 3

SEQUENTIAL RAS ADMITTING OPTIMAL NONBLOCKING SUPERVISION OF POLYNOMIAL COMPLEXITY

This chapter discusses special structure for DIS-CON-RAS that facilitates the implementation of the optimal SCP of Definition 10 in computational complexity that is polynomial with respect to the size of the controlled RAS. As it was indicated in Chapter 2, the quest for such specially structured RAS sub-classes is a naturally arising research direction, in the light of the negative result of Theorem 2.1. From a practical standpoint, the identification of the aforementioned special structure can provide a set of design guidelines that, when satisfied, will result in RAS with maximum operational flexibility. Indeed, as we shall see in the subsequent developments, the presented special structure encompasses a large set of RAS configurations that arise naturally in many contemporary technological applications, and therefore, the results of this section are of considerable practical significance.

From a methodological standpoint, the identification of the aforementioned special structure will be based on two important observations:

Observation 1: The combination of Theorems 2.1 and 2.2 implies that the super-polynomial complexity of the DIS-CON-RAS safety problem is due to the difficulty of recognizing deadlock-free unsafe states. If, however, there are RAS sub-classes that can be shown to present no deadlock-free unsafe states – i.e., for these RAS, $S_d \equiv S_{\bar{s}}$ – then, Theorem 2.2 implies that *the optimal nonblocking SCP is polynomially implementable, through the one-step-lookahead policy that tests for deadlock-freedom rather than safety.*

Observation 2: The most straightforward way to resolve the RAS state safety problem is through a search-based procedure that seeks to construct a feasible sequence of process advancing events that brings all active process instances to completion; such an event sequence will be called a *(process)* *terminating (event) sequence.* In the context of this search for terminating sequences, the non-polynomial complexity of the RAS state safety problem is manifested by the need to *backtrack* to an earlier state, every time that the search for a terminating sequence reaches a deadlock. If, however, there is some special RAS structure in which the search for a terminating sequence can be performed in a *"greedy"* manner – i.e., without the need for back-tracking – then, for the resulting RAS class, the search effort is polynomially bounded by the total number of process advancements that must be executed for the completion of all the active process instances. Clearly, this last quantity is polynomially related to the RAS size; therefore, in this case, the RAS safety problem can be addressed in polynomial complexity with respect to the RAS size. The ability to provide a "greedy" search scheme for the RAS safety problem will depend, in general, on the ability to organize the search process in a way that establishes a *monotonic* increase of the resource slack capacity.

The next section addresses special RAS structure supporting optimal non-blocking SCP through Observation 1, while Section 2 discusses special RAS structure that pertains to Observation 2.

1.   Polynomial optimal nonblocking supervision due to absence of deadlock-free unsafe states

1.1   An alternative characterization of deadlock in DIS-SU-RAS

The results presented in this section pertain to the class of DIS-SU-RAS only. Definition 7, when combined with the *single-unit* structure of the considered resource allocation, implies that all resources involved in the circular dependencies underlying a deadlock state must be filled to capacity. Hence, a more convenient working definition for deadlock in the DIS-SU-RAS class is as follows:

THEOREM 3.1 *(Reveliotis, 1996) In a DIS-SU-RAS, state s is a partial deadlock, iff there exists a resource subset, $\mathcal{DR}$, such that (i) every resource in it is allocated to capacity, and (ii) every process holding a unit of these resources requires the allocation of a unit from another resource in $\mathcal{DR}$ for the execution of its next stage.*

Proof: Let $\mathcal{DP}$ define the set of process instances allocated to resources $R_i \in \mathcal{DR}$. Then, the apparent slack of every resource $R_i \in \mathcal{DR}$ w.r.t. $\mathcal{DP}$ is equal to zero, and since every process $j_k \in \mathcal{DP}$ requires one unit of some resource $R_{j_k} \in \mathcal{DR}$ for its advancement to its next stage, it follows that the process set $\mathcal{DP}$ defines a partial deadlock. Thus, the condition described in Theorem 3.1 is a sufficient condition for deadlock in DIS-SU-RAS.

To prove that it is also necessary, suppose that state s is a deadlock state, with $\mathcal{DP}$ being the set of deadlocked processes. Then, since every process $j_k \in \mathcal{DP}$ requires just one unit of the resource R_{j_k} supporting its next stage, and the apparent slack of R_{j_k} is less than the posed requirement, it follows that R_{j_k} is currently filled to capacity with processes from the set $\mathcal{DP}$. ◇

The resource allocation requests posed by a set of active process instances in a DIS-SU-RAS can be represented in a convenient and very intuitive manner through the *Resource Dependency Graph (RDG)*, $\mathcal{G}$. The vertex set $\{V_i,\ i = 1, \ldots, m\}$, of this graph is in one-to-one correspondence with resource type set $\{R_i\ i = 1, \ldots, m\}$, while each edge E_{ij} implies that (at least) one of the units of R_i is currently allocated to a process j_k requesting a unit of R_j for a potential implementation of its next processing stage. Under the RDG representation, a DIS-SU-RAS deadlock is depicted by a *closed* subgraph of the RDG in which every node V_i corresponds to a resource R_i with slack $\delta_i = 0$; such a subgraph is characterized as a *knot* in the deadlock literature.

1.2 DIS-SU-RAS with no deadlock-free unsafe states

Our first result regarding the characterization of special RAS structure with no deadlock-free unsafe states is stated and proven as follows:

LEMMA 1 *A DIS-SU-RAS with $C_i \geq 2$, $\forall i$, has no deadlock-free unsafe states.*

Proof: To prove this lemma, we shall use the characterization of partial deadlock in DIS-SU-RAS provided by Theorem 3.1. Figure 3.1 illustrates the most important points in the proof. The result is proven by contradiction. To this end, suppose that there exist unsafe states which are *not* partial deadlocks. Then, the following series of observations hold true:

(i) First, notice that, under Condition 3 of Section 2.1 in Chapter 1, the definition of state safety implies that, while reasoning about the safety of a RAS state, it is legitimate to assume that no new processes are loaded in the system during the evolution of the system operation.

(ii) Then, under Condition 2 of Section 2.1 in Chapter 1, every unsafe state unavoidably leads to a (partial) deadlock in a finite number of process advancing events. Therefore, there exists an unsafe state s_i, s.t. further advancement of any process j_l, allocated to resource R_l and requesting a (free) unit of resource R_k, leads to a partial deadlock. Let us denote the resulting state by s_j.

(iii) Since allocating one unit of capacity of R_k leads to a partial deadlock (obviously involving resource R_k), it follows, from Theorem 3.1, that this unit must be the only slack on resource R_k. Furthermore, since $C_k \geq 2$, resource R_k must be nonempty in state s_i. Let J^k denote the (nonempty) set of processes allocated to R_k in state s_i.

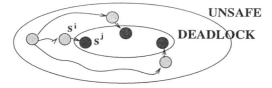

(a) The topological relationship of the Deadlock and Unsafe spaces

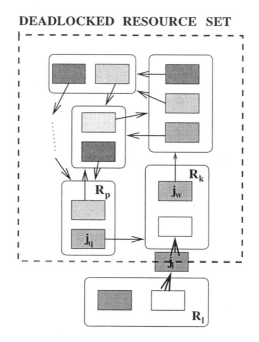

(b) The topology of the newly created partial deadlock

Figure 3.1.   Illustration of the proof of Lemma 1

(iv) Let $\mathcal{G}_i$ denote the RDG corresponding to state s_i, and consider its paths corresponding to the strings of requests constructed by starting with the resource requests posed by processes in J^k, then appending the requests of processes allocated to resources included in the first set of resource requests, and so on. Let us denote the set of resources reached through these paths – or equivalently, appearing in these strings of requests – by $\mathcal{DR}^*$. Since state s_j is a partial deadlock involving R_k, it follows that all resources in $\mathcal{DR}^*$ are allocated to capacity in s_j. Therefore, $R_l \notin \mathcal{DR}^*$.

(v) On the other hand, $R_k \in \mathcal{DR}^*$, since, otherwise, the blocking of process instances in J^k at state s_j, combined with the fact that the slack δ_k was the only resource slack reduced during the transition $s_i \to s_j$, implies that the set $\mathcal{DR}^*$ must have contained a deadlock in state s_i, as well (and contradicts the assumption that s_i is deadlock-free – cf. step (ii)).

(vi) Therefore, there exists a resource $R_p \in \mathcal{DR}^*$, with $R_p \neq R_l$, containing a process j_q requiring a unit of resource R_k. Moreover, allocating the slack unit of R_k to j_q instead of j_l, will allow a whole string of processes in $\mathcal{DR}^*$ to proceed by one step. Specifically, this string is nonempty, since it contains at least one process, j_w, from the nonempty set J^k (cf. step (iii)).

(vii) To summarize the previous steps, we started by assuming the existence of unsafe states in the system operation, which further implies the existence of (at least one) deadlock-free state s_i, in which there is no process j_l that can be advanced to its next stage without leading to the development of a partial deadlock (cf. step (ii)). Then, we established that for any process j_l that can be advanced to its next required resource R_k, (giving, thus, rise to a partial deadlock), there must exist another process j_q requiring a unit of resource R_k (cf. steps (iii)-(v)). Finally, we observed that allocating the slack unit of R_k to process j_q instead of process j_l, guarantees the feasibility of advancement of at least one process, j_w, allocated to R_k (cf. step (vi)). Since the advancement of process j_q decreases the slack of resource R_k only, and this resource is not involved in a deadlock in the resulting state s_j, it follows that s_j is deadlock-free. But this contradicts the assumption defining state s_i, and establishes the truth of Lemma 1. $\diamond$

The practical implications of this result are very powerful. For instance, in the context of the buffer space allocation for the robotic cell of Example 1 in the opening chapter, Lemma 1 implies that the maximally permissive nonblocking SCP can be polynomially implemented through one-step-lookahead for dead-lock states, as long as the local buffer at every workstation can accommodate at least two parts. This is a quite viable proposition in the context of the contemporary flexibly automated production systems. The next result indicates that similar effects can be obtained by imposing some additional structure on the underlying process flows rather than on the available resource capacities. To state this result, we need to introduce the following concepts:

the predecessor resources of R_i, $^{\bullet}R_i \equiv \{R_u : R_u$ immediately precedes R_i in some process flow $\mathcal{G}_j\}$.

the successor resources of R_i, $R_i^{\bullet} \equiv \{R_v : R_v$ immediately follows R_i in some process flow $\mathcal{G}_j\}$.

The result itself is stated as follows:

LEMMA 2 *A DIS-SU-RAS where* $|{}^\bullet R_i| = 1$ *or* $|R_i^\bullet| = 1$, *for all* R_i, *has no deadlock-free unsafe states.*

Proof: As in the proof of Lemma 1, suppose that there are deadlock-free unsafe states, and consider a state s_i which is one step away from (partial) deadlock. Let J^k denote a set of processes requesting available resource type R_k in s_i, and $\mathcal{R}(J^k)$ denote the set of resource types held by these processes in s_i. Also, let s_j denote the state resulting from the advancement of any process $j_l \in J^k$ to R_k, and $\mathcal{G}_j$ denote the corresponding RDG. We consider two cases:

Case 1 $-|{}^\bullet R_k| = 1$: Suppose that ${}^\bullet R_k = \{R_v\}$. This implies that $\mathcal{R}(J^k) = \{R_v\}$, and thus, every $j_q \in J^k$ holds an instance of R_v. But then, the generation of state s_j through the advancement of $j_l \in J^k$ implies that $\delta_v(s_j) > 0$. Furthermore, since the allocation of R_k to j_l in s_i leads to the formation of a deadlock in s_j, there must be a cycle in $\mathcal{G}_j$ from R_k to itself, R_v must belong to that cycle (since ${}^\bullet R_k = \{R_v\}$), and $\delta_v(s_j) = 0$. But this last statement contradicts the previous finding that $\delta_v(s_j) > 0$ and establishes the truth of Lemma 2 for this case.

Case 2 $-|R_k^\bullet| = 1$: Suppose that $R_k^\bullet = \{R_w\}$. Since the allocation of R_k to $j_l \in J^k$ at state s_i results in a deadlock at state s_j, we can conclude that R_w is allocated to capacity in s_i. Let J^w denote the set of processes allocated to R_w in s_i and $\mathcal{DR}^*$ be the resource set constructed by tracing their resource dependencies in $\mathcal{G}_j$, as in the proof of Lemma 1. Then, $R_k \in \mathcal{DR}^*$ (otherwise, s_i already contains a deadlock). We also claim that in s_i, $\mathcal{R}(J^k) \cap \mathcal{DR}^* = \emptyset$; indeed, in the opposite case, the advancement of a process instance $j_q \subset \mathcal{R}(J^k) \cap \mathcal{DR}^*$ to R_k will facilitate the advancement of a process in R_w, which further enables the advancement of some process in R_k, and therefore, s_j is not a deadlock state. Finally notice that $\mathcal{R}(J^k) \cap \mathcal{DR}^* = \emptyset$ in s_i contradicts the fact that $R_k \in \mathcal{DR}^*$ in s_j (since R_k can be reached in $\mathcal{G}_j$ only through the resources allocated to processes requesting a unit of R_k in s_j, and, by the construction of s_j, this resource set is a subset of $\mathcal{R}(J^k)$). The generated contradiction concludes the case and the proof. ◇

It is interesting to notice that the basic structure of the argument in the proof of Lemma 2 is identical to that of the argument in the proof of Lemma 1, and it constitutes a generic mechanism for proving this type of results. Also, in both cases, the provided proofs essentially establish that allocation of a unit from a resource type presenting the stipulated properties can be done in a way that preserves deadlock-free operation. Hence, one can combine the results of Lemmas 1 and 2 in the following theorem, that covers a broader class of DIS-SU-RAS.

THEOREM 3.2 *A DIS-SU-RAS where every resource* R_i *has either* $C_i \geq 2$ *or* $|{}^\bullet R_i| = 1$ *or* $|R_i^\bullet| = 1$, *has no deadlock-free unsafe states.*

An interesting application of Theorem 3.2 in the manufacturing domain concerns the case where every process instance *must* visit a *"central buffer"* between the execution of two consecutive processing steps; such an operational scheme has been characterized as *requisite central buffering* (Lawley and Reveliotis, 2001).

COROLLARY 1 *A DIS-SU-RAS with requisite central buffering, and at least two units of capacity at the central buffer, has no deadlock-free unsafe states.*

As a more concrete example of requisite central buffering, we notice that in many production shop-floors, jobs are transferred among the system workstations by an interconnecting conveyor loop; hence, in this case, the material handling equipment itself functions as a requisite central buffer. Another mode for engaging a central buffer in contemporary manufacturing operations is that known as *optional central buffering*. Under this scheme, a job instance having completed the execution of its current processing stage, can be routed either to its next requested workstation *or* to the central buffer. As a more concrete example, in contemporary semiconductor manufacturing systems, wafer cassettes are routed to the bay stocker if the next requested process tool is not readily available. For DIS-SU-RAS with optional central buffering, the following result is proven in (Lawley, 1999):

THEOREM 3.3 *DIS-SU-RAS with optional central buffering have no deadlock-free unsafe states.*

The result of Theorem 3.3 can be intuitively explained by noticing that, in the SU-RAS context, any free unit of capacity in the central buffer essentially provides a *swapping* mechanism for resolving any circular dependencies developing among the remaining resource types. More specifically, any two processes j_1 and j_2 situated respectively to the fully allocated resources R_1 and R_2, can be *swapped* by first transferring j_1 to the central buffer, subsequently transferring j_2 to R_1, and finally transferring j_1 to R_2. It should be obvious, then, that such a system can enter its unsafe region only through the careless allocation of the last free unit of capacity at the central buffer in a way that entangles central buffer into a deadlock.

1.3   A generalizing result

An alternative way to develop the results of Section 1.2 is by (i) trying to formally characterize the structure of deadlock-free unsafe states one step away from deadlock, and (ii) showing that the special RAS structure identified in the previous section is in conflict with the requirements for the formation of such deadlock-free unsafe states. A characterization for deadlock-free unsafe states one step away from deadlock, appropriate for the more constrained class of

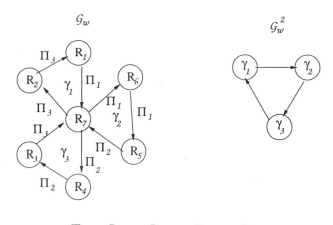

$$\Pi_1: \quad R_1 \rightarrow R_7 \rightarrow R_6 \rightarrow R_5$$
$$\Pi_2: \quad R_5 \rightarrow R_7 \rightarrow R_4 \rightarrow R_3$$
$$\Pi_3: \quad R_3 \rightarrow R_7 \rightarrow R_2 \rightarrow R_1$$

Figure 3.2.   The $\mathcal{G}_w$ and $\mathcal{G}_w^2$ graphs through an example

LIN-SU-RAS, has been developed in (Fanti et al., 1997; Fanti et al., 1998b). Next we outline the results of (Fanti et al., 1997; Fanti et al., 1998b) and we show that they provide a unifying framework for all the results reported in the previous section, when restricted to the class of LIN-SU-RAS. We also utilize these results in order to provide an alternative proof for the absence of deadlock-free unsafe states in another sub-class of LIN-SU-RAS, originally identified in (Lawley and Reveliotis, 2001).

The characterization of the deadlock-free unsafe state one step away from deadlock, provided in (Fanti et al., 1997; Fanti et al., 1998b), employs the concepts of the *working procedure digraph (WPD)*, $\mathcal{G}_w$, and its derivative graph $\mathcal{G}_w^2$.

DEFINITION 15 *The* working procedure digraph (WPD), $\mathcal{G}_w$, *of any given LIN-SU-RAS* Φ, *is a directed graph with its node set*, V_w, *being in one-to-one correspondence with the resource type set* $\mathcal{R}$, *and its edge set* E_w *consisting of all pairs* (i, j), $i, j \in \{1, \ldots, m\}$, *such that* R_j *immediately follows* R_i *in some process type* Π_j.

$\mathcal{G}_w^2$ is a directed graph that characterizes the development of a particular structure among the cycles of $\mathcal{G}_w$; a cycle γ of $\mathcal{G}_w$ is defined as a (simple) path in it with the same starting and finishing node. The node set V_w^2 of $\mathcal{G}_w^2$ consists of all the *cycles* of $\mathcal{G}_w$. An edge (γ_i, γ_j) is present in $\mathcal{G}_w^2$ *iff* (i) the two cycles intersect at exactly one resource, say R_v, and (ii) some process route contains a resource sequence $R_u \to R_v \to R_w$, where $R_u \in \gamma_i$ and $R_w \in \gamma_j$. A last concept that is necessary for the statement of the key result presented in (Fanti et al., 1997; Fanti et al., 1998b) is that of a *critical cycle* in $\mathcal{G}_w^2$: A cycle $\gamma^2 = < \gamma_1, \gamma_2, \ldots, \gamma_v >$ of $\mathcal{G}_w^2$ is critical, *iff* (i) every pair of cycles in γ^2 intersects at exactly one resource, and (ii) this intersecting resource is common for all cycle pairs, i.e., $|\gamma_1 \cap \gamma_2 \cap \ldots \cap \gamma_v| = 1$. Let $R(\gamma^2)$ denote the intersecting resource. The situation is illustrated in Figure 3.2.

THEOREM 3.4 *(Fanti et al., 1998b) A necessary condition for the existence of deadlock-free unsafe states one step away from deadlock in a given LIN-SU-RAS Φ is that (i) the corresponding digraph $\mathcal{G}_w^2$ has a critical cycle γ^2, and (ii) the intersecting resource $R(\gamma^2)$ has single capacity.*

The complete formal proof for Theorem 3.4 is provided in (Fanti et al., 1998b). Here we provide a more intuitive explanation of the result. Essentially, the deadlock-free unsafe states one step away from deadlock in LIN-SU-RAS consist of cycles in a critical cycle $\gamma^2 = < \gamma_1, \gamma_2, \ldots, \gamma_v >$ filled to capacity with process instances, except for the intersecting resource $R(\gamma^2)$. The advancement of a process instance j_i from cycle γ_i, $i = 1, \ldots, v - 1$, to $R(\gamma^2)$ will establish a deadlock involving the cycle γ_{i+1}. Similarly, the advancement of a process instance j_v from cycle γ_v to $R(\gamma^2)$ will establish a deadlock involving the cycle γ_1. The necessity of such a formation becomes obvious when considering that (i) deadlocks in LIN-SU-RAS involve at least one cycle of resources filled to capacity, and (ii) the set of cycles in $\mathcal{G}_w$ is finite. The requirement that any pair of cycles γ_i, γ_j, $i, j = 1 \ldots, v$, $i \neq j$, do not intersect at an additional resource other than $R(\gamma^2)$ is necessary since, otherwise, an argument similar to that in the proofs of Lemmas 1 and 2 would establish that $R(\gamma^2)$ could be allocated to a process instance in γ_i or γ_j without creating a deadlock. Finally, the fact that resource $R(\gamma^2)$ must be of single capacity results from the proof of Lemma 1.

We emphasize that the condition of Theorem 3.4 is necessary for the existence of deadlock-free unsafe states one step away from deadlock, but not sufficient. The point is that the test stated in this theorem can identify potential deadlock-free unsafe states one step away from deadlock, which, however, are unreachable under normal operation. Figure 3.3 exemplifies this case. The graphs $\mathcal{G}_w$ and $\mathcal{G}_w^2$ for the LIN-SU-RAS of Figure 3.3(a) are depicted in Figure 3.3(b). Notice that $\mathcal{G}_w^2$ contains the critical cycle $< \gamma_2, \gamma_3 >$ that corresponds to the deadlock-free unsafe state one step away from deadlock of Figure 3.3(c). However, the depicted state is unreachable when the LIN-SU-RAS of Figure 3.3(a) starts from the empty state and operates according to the assumptions stated in Chapter 2.

Clearly, the results of Lemmas 1 and 2, when restricted to LIN-SU-RAS, can be immediately derived from Theorem 3.4: the special structure advocated

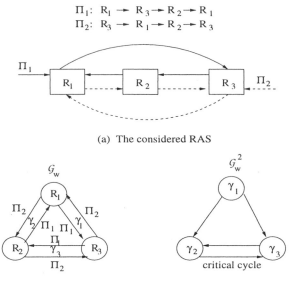

(a) The considered RAS

(b) The corresponding digraphs

(c) The unreachable state corresponding to the critical cycle

Figure 3.3.   An example of a critical cycle corresponding to an unreachable deadlock-free unsafe state

in Lemma 1 negates Condition (ii) in Theorem 3.4, while the special structure advocated in Lemma 2 negates Condition (i) in Theorem 3.4, i.e., the formation of critical cycles in $\mathcal{G}_w^2$. Next we utilize the result of Theorem 3.4 in order to prove the absence of deadlock-free unsafe states from another sub-class of LIN-SU-RAS. To formally state this result we need to introduce the concept of *"reentrant restriction"*.

DEFINITION 16 *A LIN-SU-RAS satisfies the* reentrant restriction *iff there exists an ordering for its resources,* $< R_{[1]}, R_{[2]}, \ldots, R_{[m]} >$, *such that*

$$\forall i = 1, \ldots, m, \ R_{[i]}^{\bullet} \subseteq \{R_{[i+1]}\} \cup \{R_{[j]} : j < i\}$$

In words, a LIN-SU-RAS satisfies the reentrant restriction if resources can be ordered so that any process holding some resource $R_{[i]}$ requests for its next operation either the next higher resource in the ordering, or a resource of lower order than $R_{[i]}$. We will say that process j_j requires a *"right move"* if it holds $R_{[i]}$ and requests $R_{[i+1]}$, and that it requests a *"left move"*, otherwise.

THEOREM 3.5 *(Lawley and Reveliotis, 2001) A LIN-SU-RAS with reentrant restriction where, in addition, every left move is followed by at least one right move, has no deadlock-free unsafe states.*

Proof: We shall prove this result by showing that the digraph $\mathcal{G}_w^2$ for such a LIN-SU-RAS, has no critical cycles. For this, we use contradiction. Hence, suppose that $\gamma^2 = <\gamma_1, \gamma_2, \ldots, \gamma_\nu>$ is a critical cycle of $\mathcal{G}_w^2$ with all cycles γ_k, $k = 1, \ldots, \nu$, intersecting at a single resource $R_{[i]}$. Then, for any pair $\{\gamma_k, \gamma_j\}$ with $k, j \in \{1, \ldots, \nu\}$ and $k \neq j$, the following remarks hold true:

(i) It is not possible that both γ_k and γ_j leave resource $R_{[i]}$ through a right move, since in that case, the reentrant restriction implies that resource $R_{[i+1]}$ is also an intersecting resource for these two cycles.

(ii) Similarly, it is not possible that both γ_k and γ_j leave resource $R_{[i]}$ through a left move, since in that case, the reentrant restriction implies that resource $R_{[i-1]}$ is also an intersecting resource for these two cycles.

(iii) But then, the combination of remarks (i) and (ii) above implies that $\nu = 2$. Without loss of generality, let us assume that γ_1 leaves $R_{[i]}$ through a right move and γ_2 leaves $R_{[i]}$ through a left move. The presence of edge (γ_1, γ_2) in $\mathcal{G}_w^2$ implies that there exists a process type that possesses the allocation subsequence $R_u \to R_{[i]} \to R_w$, with $(R_u, R_{[i]})$ being an edge on γ_1 and $(R_{[i]}, R_w)$ being an edge on γ_2. Since, by the working hypothesis, $(R_{[i]}, R_w)$ is a left move, $(R_u, R_{[i]})$ must be a right move. Therefore, $R_u \equiv R_{[i-1]}$, which further implies that $R_{[i-1]} \in \gamma_1$. Furthermore, the fact that γ_2 leaves $R_{[i]}$ through a left move, when combined with the reentrant restriction, imply that $R_{[i-1]} \in \gamma_2$. Hence, $R_{[i-1]} \in \gamma_1 \cap \gamma_2$, which contradicts the working hypothesis that $\gamma_1 \cap \gamma_2 = \{R_{[i]}\}$, and proves the result. ◇

2. Polynomial optimal nonblocking supervision due to "greedy" search for a process terminating event sequence

The material of this section is a generalization of some results originally presented in (Araki et al., 1977). The considered RAS class is a sub-class of the DIS-CON RAS, that is formally characterized by the following two conditions:

CONDITION 5 *For every pair of processing stages Ξ_{jk}, Ξ_{jh} such that stage Ξ_{jh} is an immediate successor of stage Ξ_{jk}, either $A_{jk}(i) \leq A_{jh}(i)$ or $A_{jk}(i) \geq A_{jh}(i)$, for all $i = 1, \ldots, |\mathcal{R}|$, i.e., every process transition corresponds to a* pure *resource allocation or deallocation. In addition, allocations and deallocations are* matched *in every processing stage sequence materializing process type* Π_j, *i.e., a set of resources allocated to* Π_j *as a block is also deallocated from it as a block.*

To introduce the second condition, we must first define the concept of *scope* of an allocation taking place in a DIS-CON-RAS satisfying Condition 5.

DEFINITION 17 *Consider a DIS-CON-RAS satisfying Condition 5. Then, a* scope *for a resource allocation to process* Π_j *is defined as a sequence* $\Xi_{jk}, \Xi_{j,k+1}, \ldots, \Xi_{j,k+\nu-1}, \Xi_{j,k+\nu}$ *of consecutive processing stages – or equivalently, a directed path in* $\mathcal{G}_j$, *the digraph representing the sequential execution logic for process type* Π_j – *such that the allocated resource units are held by* Π_j *at stages* $\Xi_{j,k+h}$, $h = 0, \ldots, \nu - 1$, *but not at stage* $\Xi_{j,k+\nu}$ *or at any immediate predecessor of stage* Ξ_{jk}. *Stages* Ξ_{jk} *and* $\Xi_{j,k+\nu}$ *are respectively characterized as the* request *and* release *stages for the considered allocation.*

The second condition characterizing the considered RAS class establishes a *"nesting"* structure for the resource allocation scopes.

CONDITION 6 *Consider a DIS-CON-RAS satisfying Condition 5 and an arbitrary execution sequence for some process* Π_j. *Then, any pair of allocation scopes defined with respect to this sequence are disjoint or one of them includes the other.*

The key result of this section is stated as follows:

THEOREM 3.6 *In the class of DIS-CON-RAS satisfying Conditions 5 and 6, safety is polynomially decidable with respect to the RAS size.*

Proof: This result is proven by providing an algorithm that resolves safety in the considered RAS class, in polynomial complexity with respect to the RAS size. To proceed with the development of this algorithm, consider a state s of a DIS-CON-RAS satisfying Conditions 5 and 6, and let $\mathcal{S}^a(s)$ denote the set of *active* process stages Ξ_{jk}, i.e., $\Xi_{jk} \in \mathcal{S}^a(s)$ iff $s(q(j,k)) \neq 0$. Then, we make the following observations:

1 The set of resource units, A_{jk}, held by any active instance j_j of some process stage $\Xi_{jk} \in \mathcal{S}^a(s)$, can be partitioned in a sequence of allocations $\mathcal{A}_{jk}^1, \mathcal{A}_{jk}^2, \ldots, \mathcal{A}_{jk}^{v(j,k)}$, such that $\mathcal{A}_{jk}^1$ took place last, $\mathcal{A}_{jk}^2$ next to last, and so on. Then, due to the nesting structure implied by Condition 6, the resource units associated with allocation $\mathcal{A}_{jk}^1$ are to be returned by the process instance j_j before the resource units associated with allocation $\mathcal{A}_{jk}^2$, which, in turn, are to be returned before the resource units associated with allocation $\mathcal{A}_{jk}^3$, etc.

2 Allocation $\mathcal{A}_{jk}^1$ is characterized as a *returnable* allocation in state s, *iff* it has a scope in $\mathcal{G}_j$ that contains Ξ_{jk} and its remaining part is executable with the resource slack capacities, $\delta_i(s)$, $i = 1, \ldots, |\mathcal{R}|$, at state s. The termination of a returnable allocation $\mathcal{A}_{jk}^1$ advances the RAS state to a state s' with $\delta_i(s') \geq \delta_i(s)$, $\forall i$; this results from the nested structure of the resource allocations stipulated by Condition 6.

3 The non-decrease of the resource slack capacities in the state s' resulting from the termination of a returnable allocation, subsequently implies that state s will be safe *iff* state s' is safe. Hence, a safety test for s can be obtained by a *greedy* algorithm that (i) iteratively seeks to identify and terminate a returnable allocation, and (ii) asserts the safety of s *iff*, in the resulting state sequence $s_{[0]}(\equiv s), s_{[1]}, s_{[2]}, \ldots, s_{[l]}$, all the allocations that are active in state s have been terminated in state $s_{[l]}$. This algorithm is detailed in Figure 3.4.

4 The number of allocations to be terminated by the algorithm of Figure 3.4 is $O(|\bigcup_j \mathcal{S}_j|)$. Since each external iteration of the algorithm involves the termination of a returnable allocation, the total number of allocation return-ability tests performed by this algorithms is $O(|\bigcup_j \mathcal{S}_j|^2)$. To establish the polynomial complexity of the considered algorithm with respect to the underlying RAS size, it remains to be shown that assessing the returnability of a given resource allocation $\mathcal{A}_{jk}^1$ can be done in polynomial time with respect to the underlying RAS size.

5 The returnability of a resource allocation $\mathcal{A}_{jk}^1$ can be evaluated by considering the $\mathcal{G}_j$ subgraph induced by the terminating sequences for this allocation emanating from stage Ξ_{jk}, and searching for a terminating sequence in it that can be supported by the resource slacks $\delta_i(s)$. This test can be implemented by a *labelling* procedure that,

    (i) starting from stage Ξ_{jk}, labels in the aforementioned $\mathcal{G}_j$ subgraph every (unlabelled) stage reachable from it under the running resource slack capacity,

    (ii) computes the adjusted resource slack vectors reflecting the advancement of an instance of process type Π_j from stage Ξ_{jk} to the newly labelled stages, and

    (iii) repeats steps (i) and (ii) on all labelled stages, until either a release stage for the considered allocation is reached or no further stages can be labelled.

The size of the $\mathcal{G}_j$ subgraph considered in the above procedure is $O(\max_j \{|\mathcal{S}_j|\})$; hence, the number of labelling tests to be performed is $O((\max_j$

State Safety Resolution Algorithm for DIS-CON-RAS satisfying Conditions 5 and 6
Input: DIS-CON-RAS $\Phi =< \mathcal{R}, C, \mathcal{P}, \mathcal{A}, \mathcal{T} >$ satisfying Conditions 5 and 6 and a state s
Output: Boolean variable SAFE

begin
   /* Initialize */
   $\forall i, \; \delta_i(s) := C_i - \sum_{q=1}^{D} s(q(j,k)) \cdot A_{jk}(i)$;
   $S^a(s) := \{\Xi_{jk}| \; s(q(j,k)) \neq 0\}$;
   $Q := \{\mathcal{A}_{jk}^1| \; \Xi_{jk} \in S^a(s)\}$;

   /* processing step */
   while ($\exists$ returnable allocations in Q) **do**
   begin
      let $\mathcal{A}_{jk}^h$ be one of them;
      remove $\mathcal{A}_{jk}^h$ from Q;
      add the resource units associated with allocation $\mathcal{A}_{jk}^h$,
      multiplied by $s(q(j,k))$, to the corresponding $\delta_i(s)$;
      if (all the allocations associated with processing stage Ξ_{jk}
         have been terminated) **then**
         remove Ξ_{jk} from $S^a(s)$;
      else
         add allocation $\mathcal{A}_{jk}^{h+1}$ to Q;
   endwhile

   if($S^a(s) = \emptyset$) **then**
      SAFE:=TRUE;
   else
      SAFE:=FALSE;
   return(SAFE);
end

Figure 3.4. A polynomial state safety resolution algorithm for DIS-CON-RAS satisfying Conditions 5 and 6

$\{|S_j|\})^2)$. Each labelling test is of complexity $O(|\mathcal{R}|)$. Therefore, the entire test is of complexity $O((\max_j\{|S_j|\})^2 \cdot |\mathcal{R}|)$. $\diamond$

We conclude this section by noticing that by modifying the labelling procedure provided in the last step of the proof of Theorem 3.6 so that it accepts only in the case that *all* the release stages of the considered allocation have been labelled, we obtain a polynomial algorithm for resolving safety in the considered RAS class, under the further assumption of uncontrollable process routings, i.e., Type-2 uncontrollability in the terminology introduced in Section 5.1 of Chapter 2.

3. Historical and bibliographical notes

To the best of our knowledge, the first formal results regarding the complexity of deadlock avoidance in sequential RAS appeared in (Araki et al., 1977; Gold, 1978). Both of these papers establish the NP-completeness of the RAS safety problem for the general case, and subsequently, they seek to identify special structure for which safety is polynomially decidable. In both cases, the identified special structure pertains to Observation 2 given in the opening discussion of this chapter, i.e., it enables a greedy search for process terminating event sequences. The relevant results developed in (Araki et al., 1977) were covered in Section 2; in particular, Theorem 3.6 is an extension of Theorem 1 in (Araki et al., 1977). However, the results presented in (Gold, 1978) were not covered in this chapter, since the dynamics underlying the resource allocation considered in that work are not exactly the same with the RAS dynamics studied in this book.

Results pertaining to the identification of special structure with no dead-lock-free unsafe states first appeared in (Xing et al., 1996; Reveliotis et al., 1997; Fanti et al., 1997; Fanti et al., 1998b), and subsequently they were classified and further investigated in (Lawley and Reveliotis, 2001). More specifically, the result of Lemma 1 was first developed independently in (Xing et al., 1996) and (Reveliotis et al., 1997), while the result of Lemma 2 together with the material of Section 1.3 first appeared in (Fanti et al., 1997; Fanti et al., 1998b). Central buffering was extensively investigated in (Lawley, 1999), while some initial observations regarding the ability of optional central buffering to function as a swapping mechanism were provided in (Wysk et al., 1994). Theorem 3.5 was first proven in (Lawley and Reveliotis, 2001) but the proof presented in Section 1.3 is new.

Chapter 4

POLYNOMIAL-KERNEL NONBLOCKING
SUPERVISORY CONTROL POLICIES
FOR SINGLE-UNIT RAS

In this chapter we take on the design of *Polynomial Kernel (PK)* SCP's that provide effective logical control for RAS not belonging to any of the classes identified in Chapter 3. We remind the reader that PK-SCP's were introduced in Chapter 2 as one-step-lookahead policies based on a condition $\mathcal{H}(s)$, $s \in S$, such that: (i)$\mathcal{H}(s_0) = \text{TRUE}$, (ii) $\mathcal{H}(s) \Rightarrow \text{safe}(s), \forall s \in S_r$, (iii) $\mathcal{H}()$ is *polynomially* testable on the system states, and (iv) the resulting SCP $\Delta_{\mathcal{H}}$ is correct, i.e., the subspace $S_r(\Delta_{\mathcal{H}})$ is strongly connected. In this chapter, we consider PK-SCP's that are appropriate for *Single-Unit* resource allocation. The derived results are generalized and extended to other RAS classes in Chapter 5.

The conditions $\mathcal{H}()$ that will be considered in both chapters, 4 and 5, will belong to one of the following two types:

Algebraic: In this case, $\mathcal{H}(s)$ constitutes a set of *linear inequalities* on the RAS state s, and the number of these inequalities is polynomially related to the RAS size $|\Phi|$. Hence, testing $\mathcal{H}(s)$ on any given RAS state s is a task of polynomial complexity with respect to $|\Phi|$.

"Greedy" Search: In this case, the condition $\mathcal{H}(s)$ corresponds to some property of the RAS state s that can be evaluated through a *search*-based algorithm, of complexity polynomial with respect to the RAS size $|\Phi|$.

We present PK-SCP's from both of these types that are appropriate for LIN-SU-RAS. We also introduce a variation of the DIS-SU-RAS that is appropriate for modelling the logical behavior of the AGV system introduced in the motivational example of Chapter 1, and provide a PK-SCP of the "greedy"-search type for it. The last part of this chapter discusses how the presented PK-SCP's can provide the building blocks for effectively addressing, in the SU-RAS context, the reconfiguration management issues introduced in Section 5.2 of Chapter 2.

From a methodological standpoint, the results of this chapter have been the outcome of a *"discovery process"* that started by *conjecturing* the appropriateness of certain conditions $\mathcal{H}(s)$ for establishing PK-SCP's for LIN-SU-RAS, and subsequently *proved* the validity of those conjectures by means of the formal framework and tools introduced in Chapters 2 and 3 for modelling and analyzing the (LIN-SU-)RAS behavior. Each of the aforementioned conditions $\mathcal{H}(s)$ essentially tries to capture one particular facet of the state safety concept as it pertains in the operation of LIN-SU-RAS. As a result, it is able to identify the safety of a particular subset of the target state space S_{rs}. The next theorem, however, indicates that there is a *synergistic* effect among the resulting policies, in the sense that they can be combined towards the synthesis of a potentially more permissive PK-SCP for the underlying RAS.

THEOREM 4.1 *Given two conditions $\mathcal{H}_1(s)$ and $\mathcal{H}_2(s)$ defining correct PK-SCP's for some RAS class, their disjunction, $(\mathcal{H}_1 \vee \mathcal{H}_2)(s)$, defines also a correct PK-SCP for the same RAS class. Furthermore, $S_r(\Delta_{(\mathcal{H}_1 \vee \mathcal{H}_2)}) = S_r(\Delta_{\mathcal{H}_1}) \cup S_r(\Delta_{\mathcal{H}_2})$.*

Proof: The fact that $S_r(\Delta_{(\mathcal{H}_1 \vee \mathcal{H}_2)}) = S_r(\Delta_{\mathcal{H}_1}) \cup S_r(\Delta_{\mathcal{H}_2})$ results immediately from the definition of PK-SCP's. To establish the correctness of $\Delta_{(\mathcal{H}_1 \vee \mathcal{H}_2)}$ we need to show that $S_r(\Delta_{(\mathcal{H}_1 \vee \mathcal{H}_2)}) \subseteq S_s(\Delta_{(\mathcal{H}_1 \vee \mathcal{H}_2)})$. Hence, consider a state $s \in S_r(\Delta_{(\mathcal{H}_1 \vee \mathcal{H}_2)})$. Then, $s \in S_r(\Delta_{\mathcal{H}_1})$ or $s \in S_r(\Delta_{\mathcal{H}_2})$; without loss of generality suppose that $s \in S_r(\Delta_{\mathcal{H}_1})$. The correctness of $\Delta_{\mathcal{H}_1}$ implies that there exists a terminating sequence for s admissible by $\Delta_{\mathcal{H}_1}$. But then, this sequence will also be admissible by $\Delta_{(\mathcal{H}_1 \vee \mathcal{H}_2)}$, and therefore, $s \in S_s(\Delta_{(\mathcal{H}_1 \vee \mathcal{H}_2)})$. ◇

Theorem 4.1 provides, thus, the motivation and justifies our insistence on identifying a *variety* of conditions $\mathcal{H}()$ that can lead to correct PK-SCP's; these policies are discussed next.

1.   Polynomial-Kernel SCP's for LIN-SU-RAS

In this section, we develop a series of PK-SCP's for LIN-SU-RAS. For each of the presented policies, we provide some discussion on the motivational ideas that led to its discovery, a formal characterization of the policy-defining condition $\mathcal{H}()$, and proofs for its polynomial complexity with respect to the underlying RAS size, $|\Phi|$, and for its correctness. Furthermore, following the PK-SCP classification introduced in the opening discussion of this chapter, we organize the subsequent development into two subsections, with the first presenting a set of *algebraic* PK-SCP's for LIN-SU-RAS, and the second focusing on PK-SCP's of the *"greedy"-search* type.

1.1 Algebraic PK-SCP's for LIN-SU-RAS

The Resource Upstream Neighborhood (RUN) Policy

The motivation behind RUN SCP. The motivation for the original specification of the RUN SCP came from the following two observations:

1 No deadlock would occur in a sequential RAS, if every process instance were allocated all the resources required for its entire processing, upon its loading into the system.

2 If at any point during its sojourn through the system, a process is allocated to a resource of very high – theoretically *infinite* – capacity, then, this process can be buffered indefinitely at that resource, without any adversarial effects for the rest of the system. In fact, for the purposes of structural analysis, the route of such a process can be decomposed to a number of segments, each of which is defined by two successive visits to the infinitely capacitated resource(s).

The first of the above observations suggests a very general *prevention* scheme for deadlock resolution, which, however, fails to take into consideration any existing information about the RAS structure, and therefore, it ends up being overly conservative and under-utilizing the system resources. This problem can be remedied, to a certain extent, when combining the first observation with the second one. More specifically, rather than trying to reserve all the resources requested by a particular process instance for its complete execution, we reserve only those resources that are going to be utilized by the process before any other resource of higher capacity is utilized by it. The situation is depicted in Figure 4.1. It turns out that, by systematically observing this *partial* reservation scheme as the process advances through the system, we can define a mechanism that leads to a correct PK-SCP for LIN-SU-RAS.

The policy-defining condition. In order to provide a detailed formal statement of the policy resulting from the above idea, we introduce the concept of the *resource upstream neighborhood*:

DEFINITION 18 *(Reveliotis and Ferreira, 1996) Given a LIN-SU-RAS Φ, the* upstream neighborhood *of resource R_i consists of all processing stages Ξ_{jk} that are supported by resource R_i, plus all the processing stages belonging to the* maximal *route subsequences immediately preceding each of the aforementioned Ξ_{jk}, and involving stages Ξ_{jp} with $C_{R(\Xi_{jp})} \leq C_i$. A process instance j_j is in the neighborhood of resource R_i iff its current processing stage is in the neighborhood of R_i.*

In Definition 18, $R(\Xi_{jp})$ denotes the unique resource type supporting the execution of stage Ξ_{jp}. The formal characterization of the condition $\mathcal{H}(s)$ defining the RUN SCP is as follows:

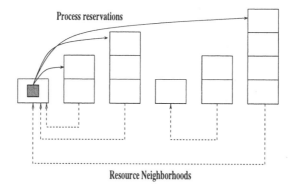

Process reservations

Resource Neighborhoods

Figure 4.1.   RUN motivation: The partial resource reservation scheme

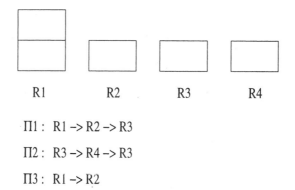

R1   R2   R3   R4

Π1 : R1 –> R2 –> R3

Π2 : R3 –> R4 –> R3

Π3 : R1 –> R2

Figure 4.2.   The LIN-SU-RAS employed for the demonstration of RUN and RO SCP's

DEFINITION 19 **RUN** *(Reveliotis and Ferreira, 1996) A resource allocation state s of a LIN-SU-RAS* Φ *is admissible by RUN SCP iff the number of process instances in the upstream neighborhood of each resource, R_i, does not exceed its buffering capacity, C_i.*

Example.   We highlight the RUN-defining logic and the resource neighborhood construction through an example. Consider the small LIN-SU-RAS depicted in Figure 4.2. This system consists of four resources, R_1, R_2, R_3, R_4, with corresponding capacities $C_1 = 2$ and $C_2 = C_3 = C_4 = 1$. In its current configuration, the system supports the production of three distinct process types, with the following routes: $\Pi_1 : R_1 \rightarrow R_2 \rightarrow R_3$, $\Pi_2 : R_3 \rightarrow R_4 \rightarrow R_3$, $\Pi_3 : R_1 \rightarrow R_2$. By applying the logic of Definition 18 to this system, we obtain the neighborhood inclusions indicated by the following incidence matrix:

$$A_{RUN} = \begin{pmatrix} 1 & & & & & & 1 & \\ & 1 & & & & & & 1 \\ & 1 & 1 & 1 & 1 & 1 & & \\ & & & & & 1 & 1 & \end{pmatrix} \qquad (4.1)$$

Each row of the above matrix corresponds to a resource neighborhood, $N(R_i)$, $i = 1 \ldots, 4$. Each column corresponds to a processing stage, starting with the stages of process type Π_1, and concatenating the stages of Π_2 and Π_3. Hence, it can be seen, for instance, that stage Ξ_{11} is in the neighborhood of resource R_1, which is required for its execution. However, it does not belong in the neighborhood of any of the resources R_2 and R_3, that are requested for the execution of the subsequent stages of process type Π_1, since the capacity of these two resources is strictly smaller than C_1. On the other hand, stage Ξ_{12} belongs in the neighborhood of R_2, which is the immediately utilized resource at that stage, but it also belongs to the neighborhood of R_3, since this resource has the same capacity with resource R_2 that supports the execution of stage Ξ_{12}, and there is no intermediate stage between stages Ξ_{12} and Ξ_{13} utilizing a higher capacity resource.

Let $C = (2, 1, 1, 1)^T$. Then, the matrix-based characterization of the RAS neighborhoods provided in Equation 4.1, when combined with the definition of the system *state* provided in Section 1.2 of Chapter 2, allows the characterization of the policy constraints implied by Definition 19, through the following system of linear inequalities on the system state:

$$A_{RUN} \cdot s \leq C \qquad (4.2)$$

◇

Complexity analysis of RUN SCP. Clearly, RUN is of polynomial complexity with respect to the RAS size, $|\Phi|$, since evaluating the policy-admissibility of any give RAS state s, requires the verification of $|\mathcal{R}|$ linear inequalities in the D system state variables. In fact, it should be easy to see that even the construction of the neighborhood sets for a single process type is of complexity no higher than $O(L^2)$, where L is the length of the corresponding route. Therefore, computing the complete incidence matrix, A_{RUN}, for the resource upstream neighborhoods, is of complexity not higher than $O(|\mathcal{P}|\bar{L}^2)$, where $\bar{L}$ denotes the length of the longest route supported by the system.

Proving the correctness of RUN SCP. Next we show that RUN is a correct PK-SCP for LIN-SU-RAS. Our discussion follows the development originally presented in (Reveliotis and Ferreira, 1996), and it utilizes the general result of Theorem 2.3 (c.f., Section 4 in Chapter 2).

THEOREM 4.2 *RUN is a correct PK-SCP for LIN-SU-RAS.*

Proof: Obviously, the condition of Definition 19 is satisfied by the RAS initial state, s_0. To prove the policy correctness in the light of Theorem 2.3, it suffices to show that, given a LIN-SU-RAS Φ and a state s of it that satisfies the RUN condition of Definition 19, there exists at least one RUN-admissible event in s that does not correspond to the loading of a new process instance. We prove this last statement by considering the following three cases:

Case 1: There exists a process instance that requests its unloading from the system. Clearly, this event can be executed without violating the feasibility conditions for the RAS operation expressed by Equation 2.7, or the RUN-defining constraint of Definition 19.

Case 2: There exists a process instance j_j that has completed its current stage Ξ_{jk}, and it requests its advancement to a stage Ξ_{jq} that is supported by a resource $R(\Xi_{jq})$ of capacity higher than or equal to the capacity of the currently held resource $R(\Xi_{jk})$. In this case, Ξ_{jk} is in the neighborhood of resource $R(\Xi_{jq})$, and the partial resource reservation scheme implied by Definitions 18 and 19 guarantees the existence of a free resource unit on $R(\Xi_{jq})$ that is currently reserved by process instance j_j. Hence, the advancement of j_j to Ξ_{jq} is feasible; it remains to establish that it is also admissible by the RUN SCP. For this, we need to argue that no resource will be over-allocated with respect to the reservation scheme established by RUN, upon the advancement of process instance j_j. Indeed, this last effect is guaranteed by the fact that $C_{R(\Xi_{jq})} \geq C_{R(\Xi_{jk})}$, which, according to Definition 18, further implies that Ξ_{jk} will be contained in every resource neighborhood that contains Ξ_{jq}.

Case 3: Every process instance in state s requests advancement to a processing stage supported by a resource with capacity strictly less than the capacity of its currently allocated resource. In this case, consider a resource R^* of *minimal* capacity among those currently allocated to active processes, and some process instance j_j allocated a unit of resource R^*, which has completed its current stage Ξ_{jk}. Then, based on the case definition and the selection of R^*, j_j requests advancement to a stage Ξ_{jq} supported by a resource $R(\Xi_{jq})$ with slack $\delta_{R(\Xi_{jq})}(s) = C_{R(\Xi_{jq})} > 0$. Hence, the advancement of j_j is feasible. It is also admissible by RUN, since any resource R_i that contains Ξ_{jq} in its neighborhood and has a non-empty neighborhood in s, will be of capacity $C_i \geq C_{R^*}$, and it will also contain stage Ξ_{jk} in its neighborhood (since $C_{R(\Xi_{jq})} < C_{R^*} \leq C_i$).

The proof of Theorem 4.2 concludes by noticing that the above three cases cover exhaustively all the possible states s of the considered RAS Φ. ⋄

Generalizing the RUN definition. A careful study of the proof of Theorem 4.2 will reveal that the entire argument holds true even if, in the construction of the resource neighborhoods described by Definition 18, we employ any other *(partial) ordering* $o()$ of the resource set $\mathcal{R}$, instead of the ordering that is induced by the resource capacities. Hence, RUN essentially defines an entire

A "counterflow" LIN–SU–RAS

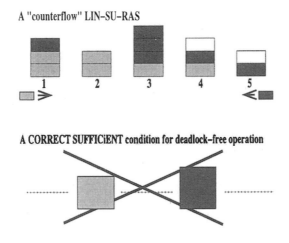

A CORRECT SUFFICiENT condition for deadlock–free operation

Figure 4.3. RO motivation: A correct sufficient condition for deadlock-free operation in "counterflow" LIN-SU-RAS

family of policies for a given LIN-SU-RAS configuration, with each member resulting from a distinct (partial) ordering of the system resources. This result becomes especially important in the light of Theorem 4.1; a systematic investigation of its implications is undertaken in Chapter 5, where we introduce a more general version of RUN, applicable to the broader class of DIS-CON-RAS.

The Resource Ordering (RO) Policy

The motivation behind RO SCP. To understand the logic behind the Resource Ordering (RO) policy, let us first concentrate on a subclass of LIN-SU-RAS, with the special property that the RAS resources can be numbered so that all process routes correspond to strictly increasing or strictly decreasing resource sequences. We characterize this particular RAS class as the *"counterflow"* LIN-SU-RAS (Lawley et al., 1997a). It is easy to see that a sufficient condition for deadlock and policy-induced deadlock-free operation in "counterflow" LIN-SU-RAS, is that no pair of resources (R_i, R_j), with $i < j$, are filled to capacity with the process instances in resource R_i corresponding to ascending resource sequences, and the process instances in resource R_j corresponding to descending resource sequences. This remark is visualized in Figure 4.3, and it is formally proven in (Lawley et al., 1997a).

Of course, the "counterflow" property is a very restrictive requirement, and a policy applicable only to this sub-class of LIN-SU-RAS would not be of any practical use. It turns out, however, that the policy-motivating idea outlined above can be extended to the more general class of LIN-SU-RAS; the solution is to *"double-count"* process instances for which the remaining route segment is non-monotonic with respect to the resource numbering.

The policy-defining condition. A complete formal characterization of RO SCP is as follows (Reveliotis, 1996; Lawley et al., 1998b):

1 Impose a *total ordering* on the set of system resources $\mathcal{R}$, i.e., apply a bijective mapping $o : \mathcal{R} \to \{1, \ldots, |\mathcal{R}|\}$ and let

$$R_i < R_j \Leftrightarrow o(R_i) < o(R_j)$$

We shall say that R_i (resp., R_j) is *"to the left"* (resp., *"to the right"*) of R_j (resp., R_i), *iff* $R_i < R_j$.

Furthermore, processing stage Ξ_{jk} is characterized as *right* (resp., *left*)-directed if $R(\Xi_{j,x-1}) <$ (resp., $>) R(\Xi_{jx})$, $\forall x > k$, where $R(\Xi_{jk})$ denotes the resource supporting stage Ξ_{jk}. A processing stage that is neither right nor left-directed is an *undirected* stage.

A process instance is characterized as *right*, *left*, or *undirected*, on the basis of its running processing stage.

2 Given a LIN-SU-RAS state s, let

$$RC_i(s) = |\{\text{right + undirected process instances in } R_i \text{ at state } s\}|$$
$$LC_i(s) = |\{\text{left + undirected process instances in } R_i \text{ at state } s\}|$$

3 Then, s is admissible by RO *iff*

$$\forall i, j : R_i < R_j \Rightarrow RC_i(s) + LC_j(s) \leq C_i + C_j - 1 \qquad (4.3)$$

Notice that by employing the total resource ordering $o()$, RO, like RUN, defines an entire *family* of SCP's for any given LIN-SU-RAS configuration; each member of this family is generated by a distinct *total* ordering imposed on the RAS resources.

Example. We elucidate the definition of RO SCP, by applying it on the small system of Figure 4.2. The ordering used in the policy implementation is the *natural* ordering of the system resources, i.e., $o(R_i) = i$, $\forall i$. Furthermore, we observe that process instances executing the last stage of their route can never deadlock the system, since their unloading from the system is always a feasible step. Hence, they can be ignored during the evaluation of the admissibility of a resource allocation state, and therefore, they are omitted during the definition of the content of $RC_i(s)$ and $LC_i(s)$.[1]

---

[1] Although not used in the example introducing the RUN SCP, a similar remark regarding the insignificance of the last process stages, applies to the formulation of all algebraic SCP's.

Table 4.1. Example: Processing stage inclusion for counters RC_i and LC_i, $i = 1, \ldots, 4$

Resource	RC_i	LC_i
R_1	Ξ_{11}, Ξ_{31}	
R_2	Ξ_{12}	
R_3	Ξ_{21}	Ξ_{21}
R_4		Ξ_{22}

It is easy to see that processing stages Ξ_{11}, Ξ_{12} and Ξ_{31} are right-directed, while processing stage Ξ_{21} is undirected, and processing stage Ξ_{22} is left-directed. Therefore, each of the counters RC_i and LC_i, $i = 1, \ldots, 4$, accumulates the instances of the processing stages presented in Table 4.1. Furthermore, part 3 of the RO Definition implies that the policy imposes the following set of linear inequalities on the system state:

$$\begin{pmatrix} 1 & & & & 1 \\ 1 & & 1 & & 1 \\ 1 & & & 1 & 1 \\ & 1 & 1 & & \\ & 1 & & 1 & \\ & & 1 & 1 & \end{pmatrix} s \leq \begin{pmatrix} 2 \\ 2 \\ 2 \\ 1 \\ 1 \\ 1 \end{pmatrix} \tag{4.4}$$

In Equation 4.4, each inequality corresponds to a pair (R_i, R_j) with $R_i < R_j$, and with all these pairs ordered lexicographically in increasing order (i.e., (R_1, R_2), (R_1, R_3), etc.). $\diamond$

Complexity analysis of RO SCP. As it was shown in the above example, the real-time implementation of RO SCP involves the testing of $O(|\mathcal{R}|^2)$ inequalities in the D state variables; therefore, the run-time complexity of the policy is polynomial with respect to the RAS size $|\Phi|$. In order to characterize the complexity of generating the policy constraints for any given LIN-SU-RAS Φ, we notice that the major task underlying this step is the characterization of its processing stages Ξ_{jk} as right, left or undirected, based on some applied resource ordering. This can be done by traversing each process route backwards, one stage at a time, and observing whether the order of the supporting resource increases or decreases at each step. As long as the observed resource ordering remains monotonically increasing (resp., decreasing), stages are characterized as left (resp., right). When the observed resource ordering undergoes a "switch", all remaining processing stages in that route are characterized as undirected. Clearly, the complexity of this task is $O(D)$, i.e., the policy can be configured in polynomial time with respect to the RAS size $|\Phi|$.

Proving the correctness of RO SCP. Next we show that RO is a correct PK-SCP for LIN-SU-RAS. The presented proof originally appeared in (Reveliotis, 1996), and similar to the RUN case, it is based on the general result of Theorem 2.3 (c.f., Section 4 in Chapter 2). We start, however, with stating and proving two results that will be used as stepping stones in the proof of the main theorem.

LEMMA 3 *Consider a state s admissible under the RO implementation on some LIN-SU-RAS Φ, and let R_k be a resource with $RC_k(s) = C_k$. Then, for every other resource $R_x > R_k$, either*

i. *R_x is not allocated to capacity at state s, or*

ii. *if it is, it contains at least one right-directed process instance.*

Proof: Suppose that R_x is allocated to capacity at state s, and it does not contain a right-directed process instance. Then $LC_x(s) = C_x$, and $RC_k(s) + LC_x(s) = C_k + C_x$ with $R_k < R_x$; this contradicts the RO-defining condition of Equation 4.3. ◇

COROLLARY 2 *Consider a state s admissible under the RO implementation on some LIN-SU-RAS Φ, and let R_k be the rightmost resource with $RC_k(s) = C_k$. Furthermore, assume that all process instances in R_k are undirected. Then, this resource is the* unique *resource with $RC_k(s) = C_k$, i.e., $\forall R_x < R_k$, $RC_x(s) < C_x$.*

Proof: Otherwise, the resource pair $< R_x, R_k >$ violates Lemma 3. ◇

Now we can state and prove the main result regarding the correctness of RO SCP.

THEOREM 4.3 *RO is a correct PK-SCP for LIN-SU-RAS.*

Proof: Obviously, the condition of Equation 4.3 is satisfied by the RAS initial state, s_0. To prove the policy correctness in the light of Theorem 2.3, it suffices to show that, given a LIN-SU-RAS Φ and a state s of it that satisfies the RO-defining condition of Equation 4.3, there exists at least one RO-admissible event in s that does not correspond to the loading of a new process instance. We prove this last statement by considering the following three cases:

Case I: There exists a *non-empty* set of resources $\mathcal{U} = \{R_u : RC_u(s) = C_u\}$. Then, the following two facts hold true:

1 $\forall R_x \in \mathcal{R}, (\exists R_u \in \mathcal{U} : R_x > R_u) \Rightarrow LC_x < C_x$

   Justification: This is an immediate consequence of Lemma 3.

2 Any right-directed process instance j_i requiring a resource R_w s.t.

- $\exists R_u \in \mathcal{U}, R_w > R_u$, and

- R_w is not filled to capacity,

can proceed.

Justification: We need to show that this step is admissible by RO. Let s' denote the RAS state resulting from the advancement of j_i to R_w. Since the considered process instance is right-directed, once it is in R_w, it affects only the RO constraints involving R_w and the resources being to the right of it. Suppose that R_x is a resource with $R_x > R_w$. Then, since by assumption $\exists R_u \in \mathcal{U}, R_w > R_u$, it follows that $R_x > R_u$, and from Fact 1, it has to be $LC_x(s) < C_x$. Thus, $RC_w(s') + LC_x(s') < C_w + C_x$ is satisfied.

Let R_k be the rightmost resource in $\mathcal{U}$, i.e., $R_k \in \mathcal{U} \wedge R_k \geq R_u, \forall u \in \mathcal{U}$. To show the existence of an admissible process-advancing event under the assumption of Case I, we further distinguish between the following two subcases:

Subcase I.1: There exists a right-directed process instance j_i in resource R_k.

  i If j_i requires a unit of resource R_l not filled to capacity, then, according to Fact 2, it can proceed.

  ii If j_i requires a unit of resource R_l filled to capacity, then, according to Lemma 3, R_l contains a right-directed process instance j_i'. Then, all the previous statements w.r.t. to process instance j_i, apply also to process instance j_i'. Since resources are finite and ordered, there exists a unique rightmost resource filled to capacity, and so, by repeating the above argument a *finite* number of times, it is possible to identify a process instance able to proceed in this subcase.

Subcase I.2: All process instances in resource R_k are undirected.

Notice that according to Corollary 2, R_k must be the unique resource with $RC_i(s) = C_i$. Without loss of generality, suppose that there exists a process instance j_i in R_k requiring a right step, to one unit of resource R_l. Then, either

  i R_l is filled to capacity, in which case the argument of Subcase I.1.(ii) applies, or

  ii R_l has free capacity, and therefore, the advancement of j_i to its next stage is feasible. This step will also be admissible by RO, since

    - for $R_x > R_l$, we have established that $LC_x(s) < C_x$ (by fact 1), and

    - for $R_x < R_l$, we have $RC_x(s) < C_x$, since, in state s, R_k was the only resource with $RC_k(s) = C_k$ (by Corollary 1).

Case II: There exists a *non-empty* resource set $\mathcal{U} = \{R_u : LC_u(s) = C_u\}$. For this case, we can apply a line of reasoning which is *dual* to that applied in Case I, in the sense that

- we are concerned with motion in the left direction,

- we change $RC(s)$ with $LC(s)$ and vice versa,

- we reverse the inequalities in resource orderings, and

- we change the characterization "rightmost" with "leftmost".

Case III: $\forall R_u \in \mathcal{R},\ LC_u(s) < C_u \wedge RC_u(s) < C_u$. For this case, first, notice that if a process j_i waits upon a resource R_w with free capacity, then j_i can proceed, since, in the resulting state s',

- $\forall R_x < R_w,\ RC_x(s) < C_x \implies RC_x(s') + LC_w(s') < C_x + C_w$

- $\forall R_x > R_w,\ LC_x(s) < C_x \implies RC_w(s') + LC_x(s') < C_w + C_x$

It remains to show that such a process instance will always exist. Suppose not. Then we have a resource subset $\mathcal{U}$ filled to capacity with process instances waiting upon resources in $\mathcal{U}$. Since resources are ordered, let $R_{u_{max(resp.,min)}}$ be the rightmost (resp., leftmost) resource in $\mathcal{U}$. Then the next step for all process instances in $R_{u_{max(resp.,min)}}$ is leftwards (resp., rightwards). But this implies that $LC_{u_{max}}(s) = C_{u_{max}}$ and $RC_{u_{min}}(s) = C_{u_{min}}$, which contradicts the case assumption. ◇

1.2   "Greedy" Search-based PK-SCP's for LIN-SU-RAS

Ordered RAS states and Banker's Algorithm

The motivation behind Banker's algorithm.   The motivating idea for the PK-SCP presented in this section stems from the fundamental Observation 2 provided in the introduction of Chapter 3, and further pursued in Section 2 of the same chapter. An important concept underlying those results was that of process-advancing event sequences structured according to a set of *"milestone"* events, such that the achievement of each milestone event guarantees a *monotonic* increase of the resource slack capacities. We saw that such sequences can be identified through a *"greedy"* search for the next milestone event, and the resulting algorithm possesses polynomial complexity with respect to the RAS size $|\Phi|$. This finding can be employed in the design of PK-SCP's for LIN-SU-RAS, by seeking to constrain the RAS operation in those states that possess such easily identifiable process-*terminating* event sequences. Clearly, the resulting policy is correct. Next, we proceed with the formalization of these general ideas.

Ordered States. The aforementioned idea of states possessing an easily identifiable process-terminating event sequence is formalized by the concept of the *ordered* RAS state.

DEFINITION 20 *(Lawley et al., 1998a) Consider a reachable state $s \neq s_0$ of a LIN-SU-RAS Φ, and let $S^a(s) = \{\Xi_{jk} : s(q(j,k)) \neq 0\}$; i.e., $S^a(s)$ is the set of processing stages with active process instances in s. Then, state s is ordered, iff there exists an ordering $o : S^a(s) \to \{1, \dots, |S^a(s)|\}$, such that an active process instance of the h-th processing stage could run to completion, given (i) the resources currently allocated to it, (ii) the system slack capacities $\delta_i(s)$, $i = 1, \dots, |\mathcal{R}|$, and (iii) the resources held by the active process instances in the first $(h - 1)$ processing stages. Furthermore, state s_0 is ordered, by convention.*

Let S_o denote the set of ordered states of Φ. Next we show that the RAS subspace induced by the state set $S_{ro} \equiv S_r \cap S_o$ is strongly connected.

THEOREM 4.4 *For any LIN-SU-RAS Φ, the subspace induced by S_{ro} is strongly connected and contains the initial state s_0.*

Proof: $s_0 \in S_r \cap S_o$, by definition of S_r and S_o. To establish the strong connectivity of S_{ro}, it suffices to show that for any state s in it, there exists a *path* leading to the RAS initial state s_0. For this, we use Theorem 2.3 (c.f. Section 4 in Chapter 2). Hence, consider an ordered state $s \in S_{ro}$ and let $o()$ denote the ordering of the set of its active processing stages $S^a(s)$, implied by Definition 20. Furthermore, let Ξ_{jk} be such that $\Xi_{jk} \in S^a(s)$ and $o(\Xi_{jk}) = 1$. Next we show that the state s' resulting from s by advancing an active processing instance in stage Ξ_{jk} by one step, belongs to S_{ro}.

The fact that $s' \in S_r$ is immediately implied by the selection of Ξ_{jk} and Definition 20. To prove that $s' \in S_o$, we discern the following six cases:

Case 1: $s(q(j,k)) = 1 \wedge k = l(j)$. Then, a valid ordering for state s' is $o'() : S^a(s)\backslash\{\Xi_{jk}\} \to \{1, \dots, |S^a(s)| - 1\}$, with $o'(\Xi_{pq}) = o(\Xi_{pq}) - 1$, $\forall \Xi_{pq} \in S^a(s)\backslash\{\Xi_{jk}\}$.

Case 2: $s(q(j,k)) = 1 \wedge k \neq l(j) \wedge s(q(j,k+1)) = 0$. Then, a valid ordering $o'()$ for state s' is obtained by setting $o'(\Xi_{j,k+1}) = 1$, and maintaining the same order for every other processing stage $\Xi_{pq} \in S^a(s)\backslash\{\Xi_{jk}\}$.

Case 3: $s(q(j,k)) = 1 \wedge k \neq l(j) \wedge s(q(j,k+1)) > 0$. Then, a valid ordering $o'()$ for state s' is obtained by setting $o'(\Xi_{j,k+1}) = 1$, $o'(\Xi_{pq}) = o(\Xi_{pq})$, $\forall \Xi_{pq} \in S^a(s)\backslash\{\Xi_{jk}\}$ with $o(\Xi_{pq}) < o(\Xi_{j,k+1})$, and $o'(\Xi_{pq}) = o(\Xi_{pq}) - 1$, $\forall \Xi_{pq} \in S^a(s)$ with $o(\Xi_{pq}) > o(\Xi_{j,k+1})$.

Case 4: $s(q(j,k)) > 1 \wedge k = l(j)$. Then, the ordering $o()$ of state s constitutes a valid ordering for s', as well.

Banker's Algorithm for LIN-SU-RAS
Input: LIN-SU-RAS Φ and a state s
Output: Boolean variable ORDERED

1 Set
   $\mathcal{S}^a := \{\Xi_{jk} : s(q(j,k)) \neq 0\}; i = 0;$ ORDERED:=TRUE.

2 Repeat

   (a) $i := i + 1;$

   (b) Try to find an active processing stage $\Xi_{jk} \in \mathcal{S}^a$, the instances of which can terminate by using their currently allocated resources, plus the currently free resource units.

   (c) If no such a processing stage can be found, ORDERED:= FALSE.

   (d) else $o(\Xi_{jk}) := i; \mathcal{S}^a := \mathcal{S}^a \backslash \{\Xi_{jk}\}$; release the resources held by the active process instances of Ξ_{jk} to the pool of the free resource units.

   until $(\mathcal{S}^a = \emptyset) \vee$ (ORDERED=FALSE)

3 return ORDERED

Figure 4.4. An implementation of Banker's algorithm for LIN-SU-RAS (Lawley et al., 1998a)

Case 5: $s(q(j,k)) > 1 \wedge k \neq l(j) \wedge s(q(j, k+1)) = 0$. Then, a valid ordering $o'()$ for state s' is obtained by setting $o'(\Xi_{j,k+1}) = 1$, $o'(\Xi_{pq}) = o(\Xi_{pq}) + 1$, $\forall \Xi_{pq} \in \mathcal{S}^a(s)$.

Case 6: $s(q(j,k)) > 1 \wedge k \neq l(j) \wedge s(q(j,k+1)) > 0$. Then, a valid ordering $o'()$ for state s' is obtained by setting $o'(\Xi_{j,k+1}) = 1$, $o'(\Xi_{pq}) = o(\Xi_{pq}) + 1$, $\forall \Xi_{pq} \in \mathcal{S}^a(s)$ with $o(\Xi_{pq}) < o(\Xi_{j,k+1})$, and $o'(\Xi_{pq}) = o(\Xi_{pq})$, $\forall \Xi_{pq} \in \mathcal{S}^a(s)$ with $o(\Xi_{pq}) > o(\Xi_{j,k+1})$.

The proof concludes by noticing that Cases 1 - 6 cover exhaustively all the possibilities for state s. $\diamond$

The LIN-SU-RAS Banker's Algorithm.   Theorem 4.4 established the correctness of the one-step-lookahead policy that accepts only transitions to ordered states. To establish that such a policy is a PK-SCP, we need to provide an algorithm that recognizes ordered states in polynomial complexity with respect to the RAS size $|\Phi|$. Such an algorithm is provided in Figure 4.4, and it is known in the literature as *Banker's* algorithm. Clearly, the number of iterations of Step 2 in the algorithm of Figure 4.4 is no more than $O(D^2)$, where D denotes the total number of processing stages in Φ. Applying the test of Step (2.b) on

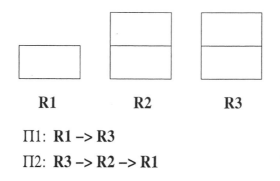

$$\Pi 1: \ \mathbf{R1} \rightarrow \mathbf{R3}$$
$$\Pi 2: \ \mathbf{R3} \rightarrow \mathbf{R2} \rightarrow \mathbf{R1}$$

Figure 4.5. The LIN-SU-RAS employed for the demonstration of Banker's algorithm and BK-DAA SCP

any given process instance Ξ_{jk} is of complexity $O(|\mathcal{R}|)$. Hence, the overall complexity of the algorithm is no more than $O(|\mathcal{R}|D^2).$[2]

Example. Consider the LIN-SU-RAS of Figure 4.14. This system consists of three resources, R_1, R_2, R_3, with capacities $C_1 = 1$ and $C_2 = C_3 = 2$. In its current configuration, the system supports two process types: $\Pi_1 : \ R_1 \rightarrow R_3$ and $\Pi_2 : \ R_3 \rightarrow R_2 \rightarrow R_1$. The reader should be able to verify that state $s = (1\,0\,1\,1\,0)^T$ is ordered, with a valid ordering being $o(\Xi_{11}) = 1, o(\Xi_{22}) = 2$ and $o(\Xi_{21}) = 3$. On the other hand, state $s' = (1\,0\,2\,1\,0)^T$ is not ordered, even though it is safe: advancing one instance of processing stage Ξ_{21} to its next stage allows the process instance in stage Ξ_{11} to run to completion, which further allows the remaining processes to finish. It is the inability of Banker's algorithm to recognize the existence of process-terminating event sequences containing such partial process advancements that renders it suboptimal. ◇

Banaszak & Krogh's Deadlock Avoidance Algorithm (BK-DAA)

Banaszak & Krogh's Deadlock Avoidance Algorithm (BK-DAA) appeared originally in (Banaszak and Krogh, 1990), and it constitutes one of the very first efforts to develop provably correct and scalable deadlock avoidance policies for sequential RAS, while exploiting the a priori available information about the process routes. The original development of (Banaszak and Krogh, 1990) employed the Petri Net modelling paradigm (Murata, 1989) for the statement and analysis of the policy and its properties. Here we shall provide a more informal

---

[2]We note that (Habermann, 1969) provides an implementation of Banker's algorithm of complexity $O(|\mathcal{R}|D \log D)$.

characterization of the policy, and we shall outline the primary mechanism that guarantees its correctness.

The motivation behind BK-DAA.   The motivational idea for BK-DAA SCP is similar to that behind Banker's algorithm, in that both policies seek to constrain the system operation to states with some easily identifiable process-terminating event sequence. However, instead of looking directly for process-terminating sequences, BK-DAA seeks to guarantee the process advancement out of their *"critical regions"*, i.e., those segments of their routes that engage resources utilized by more than one processing stage, and therefore, susceptible to deadlock. Indeed, if the system can reach a resource allocation state where all running process instances are concentrated to routing segments where every stage utilizes exclusively its supporting resource, then, the *acyclicity* of the process routes guarantees that all these processing instances can be completed through the following "pulling" scheme: Process advancements are grouped on a type by type basis, with process instances of each type being sequenced in a way that gives the highest priority to the most advanced process instances; process types themselves can be sequenced in any arbitrary order.

Shared and Unshared Resources and Production Zones.   The above idea is formalized to a policy as follows: Given a LIN-SU-RAS Φ, we characterize any resource of it as an *unshared resource* if it supports only one processing stage. Let us denote the set of unshared resources in Φ by $\mathcal{R}^u (\subseteq \mathcal{R})$. Then, $\mathcal{R}^s = \mathcal{R} \setminus \mathcal{R}^u$ defines the set of *shared resources*[3]. Furthermore, given a process type Π_j of Φ, consider the *maximal* route segments consisting of successive unshared resources, in the sense that the immediately preceding and succeeding resources in the route are either shared or they do not exist. We refer to these segments as the *unshared sub-zones* of process route Π_j. The interleaving segments of shared resources are called the *shared sub-zones* of Π_j. A pair of consecutive shared and unshared sub-zones in a process route constitutes a *production zone*, in the terminology of (Banaszak and Krogh, 1990). Hence, a process route is decomposed to a sequence of production zones, i.e. $\Pi_j = < z_j^1, z_j^2, \ldots, z_j^{l(j)} >$, $j = 1, \ldots, n$. The shared and unshared sub-zones of production zone z_j^q will be denoted respectively by s_j^q and u_j^q. It should be noticed that production zone z_j^1 (resp., $z_j^{l(j)}$) might have its shared (resp., unshared) sub-zone, s_j^1 (resp., $u_j^{l(j)}$), missing. Furthermore, $C_{u_j^q}$ will denote the total capacity of sub-zone u_j^q, i.e., $C_{u_j^q} = \sum_{\Xi_{jk} \in u_j^q} C_{R(\Xi_{jk})}$; by convention, $C_{u_j^q} = \infty$ if $u_j^q = NULL$. Finally, given a resource $R_i \in \mathcal{R}^s$ and some processing stage Ξ_{jk} supported by R_i, let

---

[3]Notice that a resource is shared even if it is only used by two or more processing stages of the same process route; this feature is necessary if BK-DAA is to be correct for LIN-SU-RAS with *re-entrant* process routes.

$\mathcal{R}^s(\Xi_{jk})$ denote the set of resources which must be allocated to a processing instance at stage Ξ_{jk} in order to complete the execution of its current shared sub-zone.

BK-DAA. In the light of the above characterizations, BK-DAA is expressed by the following two constraints, that regulate the loading and advancement of the various processing instances at any state s of a given LIN-SU-RAS Φ:

1 The total number of process instances, j_j, that are executing stages of a production zone z_j^q, must not exceed the total capacity of its unshared sub-zone, $C_{u_j^q}$.

2 The advancement of a process instance j_j to some processing stage Ξ_{jk} that belongs to a shared sub-zone s_j^q is admissible, only if every resource $R_i \in \mathcal{R}^s(\Xi_{jk})$ has a slack $\delta_i(s) > 0$, i.e., only if j_j could complete the entire route segment corresponding to sub-zone s_j^q without advancing any other process instances.

Condition 1 essentially regulates the process advancement among the different production zones, ensuring that all active processes in each production zone could be buffered, if necessary, in the corresponding unshared sub-zone. Condition 2 ensures that there will always exist a process advancing sequence that will concentrate all active process instances to the unshared sub-zones of their currently executing production zones. Specifically, for each production zone, this process advancing sequence will first advance processes in the unshared sub-zone as far as possible, making the remaining sub-zone capacity available for the processes in the corresponding shared sub-zone. Subsequently, it will advance the processes in the shared sub-zone all the way to the unshared sub-zone, giving priority to the processes that were advanced most recently; Condition 2 guarantees the feasibility and admissibility of such a process advancing sequence. In the particular case that a production zone has no unshared sub-zone, the relevant requirements are trivially met by considering the RAS environment acting as an unshared sub-zone of infinite capacity. Once all processes have been aggregated in their unshared subzones, a terminating sequence for their remaining processing steps can be constructed according to the "pulling" scheme described in the opening discussion for BK-DAA; Condition 1 will be satisfied by this scheme since processes will always be advanced towards empty production zones, while Condition 2 will be met since this advancing scheme will let at most one process instance in the RAS part consisting of its shared resources. The above remarks establish that BK-DAA is a correct policy. Finally, under the assumption that the RAS controller keeps track dynamically of the slack capacity for each resource and each unshared sub-zone, the computational complexity for testing Condition 1 is $O(1)$, while the complexity for testing Condition 2 is $O(|\mathcal{R}^s|)$.

Example. We demonstrate the BK-DAA SCP and its underlying dynamics using the example RAS of Figure 4.14. For this system, $\mathcal{R}^u = \{R_2\}$ and $\mathcal{R}^s = \{R_1, R_3\}$. Hence, process type Π_1 consists of a single production zone $u_1^1 \equiv s_1^1 = < R_1, R_3 >$, while process type Π_2 has two production zones: z_2^1 with shared sub-zone $s_2^1 = < R_3 >$ and unshared sub-zone $u_2^1 = < R_2 >$, and $z_2^2 \equiv s_2^2 = < R_1 >$. The capacity to be observed by Condition 1 with respect to production zone z_2^1 is equal to C_2, while for the remaining zones, it is infinite (since they have no unshared sub-zones).

Suppose that the considered RAS is in state $s_1 = (1\,0\,1\,0\,0)^T$ and we attempt to load a new instance of process type Π_2. This event is admissible by BK-DAA since both Conditions 1 and 2 are satisfied by the resulting state. On the other hand, if we attempt to load this new process instance while the system is in state $s_2 = (1\,0\,0\,2\,0)^T$, this event will be blocked by BK-DAA since Condition 1 will be violated for production zone z_2^1. It is interesting to notice that Banker's algorithm would make the opposite decision in each of these two cases! ◊

2.   Efficiency Considerations

The multitude of policies introduced in the previous section arises naturally the question of how to assess and compare the efficacy of two logical controllers for a given LIN-SU-RAS configuration, that result from the implementation of two different PK-SCP's on it, or, in the case of RUN and RO SCP's, even the same SCP but under two different resource orderings. The most straightforward and complete way to answer this question is by referring it to the basic description of the overall RAS supervisory control problem provided in Section 3 of Chapter 1. Hence, in the light of that discussion, the best of the two logical controllers is the one that enables a performance control policy that results in the best possible value for the performance index under consideration, among the performance control policies enabled by any of these two controllers. This kind of analysis for the policy efficacy necessitates the characterization of the system timed-based dynamics and performance, as well as the nature and structure of the applied performance control policies, and therefore, it is deferred to Chapter 6, that undertakes these performance considerations. In the rest of this section we focus on another performance index, that relates more directly to the objectives of the logical control framework introduced in Chapter 2, and the notion of optimality defined therein. We remind the reader that the optimal SCP, Δ^*, introduced in Section 2.3 of Chapter 2, is the policy that accepts the entire reachable and safe space, S_{rs}, of the considered RAS, while any PK-SCP Δ will tend to give up some of this space in order to establish polynomial "run-time" complexity. Hence, a natural measure of the policy efficiency in this context, is the ratio

$$I_\Delta \equiv \frac{|S_r(\Delta)|}{|S_{rs}|} \qquad (4.5)$$

We perceive I_Δ to be a measure of the *operational flexibility* enabled by policy Δ. Next, we provide some remarks that can enhance the operational flexibility of PK-SCP-based controllers for any given LIN-SU-RAS configuration.

2.1   Essentially Different PK-SCP's

The first remark pertains to the result of Theorem 4.1, established in the opening section of this chapter. According to Theorem 4.1, the disjunction of two PK-SCP's, Δ_1 and Δ_2, defines another PK-SCP Δ_3 with $S_r(\Delta_3) = S_r(\Delta_1) \cup S_r(\Delta_2)$. This result becomes especially important in the case that

$$S_r(\Delta_1) \nsubseteq S_r(\Delta_2) \wedge S_r(\Delta_2) \nsubseteq S_r(\Delta_1) \tag{4.6}$$

since, then, policy Δ_3 is more flexible than any of the two constituent policies. We shall characterize two policies, Δ_1 and Δ_2, that satisfy the condition of Equation 4.6 as *essentially different*. The example on the BK-DAA SCP, that was provided in Section 1.2.0 established that BK-DAA and Banker's algorithm are essentially different SCP's. Similarly constructed examples can establish the essential difference of RUN with the other three PK-SCP's introduced in Section 1, as well as the essential difference of RO with BK-DAA. On the other hand, we invite the reader to show that RO PK-SCP is subsumed by Banker's algorithm, i.e., for any given LIN-SU-RAS configuration, every state that satisfies the RO-defining condition of Equation 4.3 is ordered. We opted to maintain RO SCP in our exposition, since its algebraic structure might facilitate the easier integration of the policy-defining logic in some formulations addressing broader design issues, than the search-based characterization of the ordered states admitted by Banker's algorithm.

2.2   Optimal and Orthogonal Resource Orderings for RUN and RO SCP's

The resource ordering $o()$ employed in the definition of RUN and RO SCP's, essentially defines a *"parameter"* that must be selected during the implementation of these two policies on any given LIN-SU-RAS configuration. In the context of the policy optimization framework defined by Equation 4.5, the best ordering will be the one that leads to the *least restrictive* set of policy-defining constraints (c.f., Equations 4.2 and 4.4). The rigorous characterization of this last criterion seems to be an intractable problem; therefore, we propose to use as a "proxy" to this criterion, the *sparsity* of the matrix A that defines the left-hand-side of the policy constraints. In particular, the *"optimal" resource ordering* to be employed in the implementation of any of these two policies will be the one that minimizes the sum of the elements of the resulting matrix A. A detailed formulation of the "optimal ordering selection" problem for the case of RUN SCP is provided in Section 4.2 of Chapter 5, where the policy is generalized so that it applies to the entire family of DIS-CON-RAS. A similar formulation for the case of RO SCP is provided in (Lawley et al., 1998b).

An additional problem is the selection of the second and the subsequent resource orderings, in the case that one seeks to develop a control policy for a given LIN-SU-RAS, that constitutes a disjunction of a number of RUN or RO implementations on this RAS. The heuristical idea that we propose for driving these selections is to *differentiate as much as possible the content of the policy constraints resulting from the newly selected ordering, compared to the content of the constraint sets established by the already selected orderings.* This can be achieved by seeking to place the unit entries for the incidence matrix of the new policy instantiation to positions that were occupied by zeroes in the matrices corresponding to prior policy implementations, to the extent possible. For this reason, we characterize this criterion and the corresponding ordering selection problem as the identification of *"orthogonal" orderings* to those already selected. The formulation of the "orthogonal" ordering selection problem for the case of RUN SCP is also provided in Chapter 5, Section 4.2.

2.3   Combining PK-SCP's with Partial Search

The operational flexibility of PK-SCP's can also be enhanced by allowing transitions to states s that fail to satisfy the policy-defining condition $\mathcal{H}()$, but their inclusion in the target set S_{rs} can be established through *partial search* of some predetermined depth n; the resulting control scheme is known as *n-step look-ahead*. More specifically, state s with $\mathcal{H}(s) = \text{FALSE}$ is admissible, under n-step look-ahead, if there exists a sequence of feasible events, u, such that (i) $|u| \leq n$, and (ii) $f(s, u) = s' \in S_r(\Delta_{\mathcal{H}})$. Since this new admissibility condition is of *existential* character, it is deemed that it can increase the policy efficiency with rather small computational cost, for reasonable sizes of look-ahead horizons.

It is interesting to notice how the length, n, of the look-ahead horizon partitions the target set S_{rs} accepted by the optimal SCP, Δ^*: $n = 0$ defines the *kernel* set $S_r(\Delta_{\mathcal{H}})$, while every increase of the look-ahead horizon by one step, say from n to $n + 1$, admits an additional *"ring"* of states. Obviously, for finite state spaces, this expansion continues only up to the point that the entire set S_{rs} is covered, for some maximal length N. The partitioning of the optimal subspace on the basis of the lookahead horizon length, n, is depicted in Figure 4.6.

2.4   Decomposing the original RAS to its "critical regions"

A last observation that can lead to considerable improvement of the operational flexibility allowed by the SCP applied on any given LIN-SU-RAS configuration, is that the deployed nonblocking supervisor must regulate only

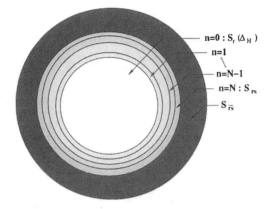

Figure 4.6. Expanding the subspace admitted by a PK-SCP through n-step lookahead

the allocation of those resource types that can be entangled in *"circular waiting"* patterns. These are the resources that belong in the strongly connected components of the RAS *Working Procedure Digraph (WPD)*, $\mathcal{G}_w$, introduced in Section 1.3 of Chapter 3; we shall characterize each strongly connected component of the graph $\mathcal{G}_w$ as a *critical region* of the corresponding RAS. Hence, the overall nonblocking supervisory control function is naturally decomposed to a number of SCP's, one for each critical region of the considered RAS. Every process route is decomposed to a number of stage subsequences alternating between those that are supported by resources belonging to some critical region of the RAS, and those that are supported by resources of naturally deadlock-free allocation. The control logic of each critical region applies only to those process sub-routes executed in that region. For instance, the partial reservation scheme corresponding to a RUN implementation on a particular critical region will be enacted for a certain process instance only upon its entrance in this region, and it will concern only resources supporting the process segment that is to be executed in this critical region. Similarly, the application of Banker'a algorithm on a particular critical region will seek to sequence only the active processing stages supported by resources in this region, and in a way that their process instances can complete only the remaining route segment in that region, rather than their entire processing through the RAS. Clearly, by reducing the span of the SCP application with respect to the RAS process routes, the afore-mentioned decomposition enhances, both, the operational and computational efficiency of the applied SCP's. In fact, the following example demonstrates that the proposed decomposition can even enable the polynomial implementation of the optimal SCP, Δ^*, for RAS that fail to meet the conditions identified in Chapter 3.

The LIN–SU–RAS

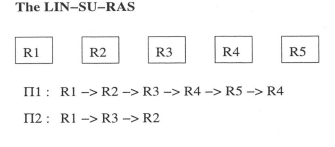

$\Pi 1:$ R1 –> R2 –> R3 –> R4 –> R5 –> R4

$\Pi 2:$ R1 –> R3 –> R2

The Working Procedure Digraph

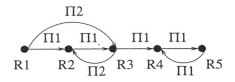

Figure 4.7.   Example: The considered RAS and its Working Procedure Digraph

Example.   Consider the LIN-SU-RAS depicted in Figure 4.7. The process routes and the WPD, $\mathcal{G}_w$, of this RAS are also depicted in Figure 4.7. $\mathcal{G}_w$ has three strongly connected components: $\{R_1\}$, $\{R_2, R_3\}$ and $\{R_4, R_5\}$. The route segments to be controlled with respect to the RAS critical region defined by the strongly connected component $\{R_2, R_3\}$, are $< \Xi_{12}, \Xi_{13} >$ and $< \Xi_{22}, \Xi_{23} >$. It should be obvious that the necessary and sufficient condition for establishing deadlock-free operation in this critical region is expressed by the following linear inequality:

$$\Xi_{12} + \Xi_{22} \leq C_2 + C_3 - 1$$

The only route segment corresponding to the RAS critical region $\{R_4, R_5\}$ is $< \Xi_{14}, \Xi_{15}, \Xi_{16} >$, and the necessary and sufficient condition for deadlock-free operation in this critical region is:

$$\Xi_{14} + \Xi_{15} \leq C_4 + C_5 - 1$$

Hence, in this case, the proposed decomposition allowed us to identify a closed-form representation of the optimal SCP, which is also of polynomial complexity with respect to the underlying RAS size, even though the considered RAS does not belong to any of the RAS sub-classes considered in Chapter 3.

3. AGV RAS and the AGV Banker's Algorithm

In this section, we consider the problem of establishing deadlock-free routing in the zoned-controlled AGV environment that was discussed in the second motivational example of Chapter 1. More specifically, we provide a detailed modelling of that operational environment as a resource allocation system, and we develop a variation of Banker's algorithm that can establish deadlock-free operation for the resulting RAS, and presents polynomial complexity with respect to the RAS size. Most of the results presented in this section have appeared originally in (Reveliotis, 2000b).

3.1 A RAS model for the considered AGV system and the study of its deadlock-related properties

The AGV RAS. The operation of the considered AGV system – c.f. the discussion in the second motivational example of Chapter 1 and Figure 1.3 – is formally abstracted to a resource allocation system as follows: The system topological structure is modelled by the AGV *guidepath graph*, $G = (N, E)$, that is assumed to be *undirected* and *connected*. The edge set E of this graph is defined by the AGV zones corresponding to the path links of the network, and, as it was explained in the discussion of Example 2, it constitutes the *resource set* of the underlying RAS model. The node set N corresponds to the intersections, workstations and the docking station of the actual AGV network, as well as the interconnecting points of any guidepath links that are artificially segmented to more than one zones. In particular, the system workstations are modelled by a set of "sink" nodes N_W connected to the rest of the network by a single edge. This edge corresponds to the operational zone for vehicles loading and unloading parts to/from the station, and it can be occupied only by vehicles trying to access the workstation. An AGV entering the edge leading to some workstation node $n_i \in N_W$, immediately changes its direction on that edge, heading back towards the rest of the network. Similarly, the docking station is modelled by another terminal node, d_s, connected to the rest of the network with a single edge. However, a vehicle entering this node, vanishes. This effect models the ability of the docking station to accommodate the entire AGV fleet, which renders it equivalent to an *infinite*-capacity resource in the AGV RAS operation. Except from these two singularities characterizing the operation of the vehicles at the system workstations and the docking station, vehicle motion on the rest of the guidepath graph is *unidirectional*, i.e., vehicles must traverse a guidepath edge from the node that they entered it to the opposite one, and they cannot "back up" in it. The guidepath graph of the AGV layout depicted in Figure 1.3 is given in Figure 4.8.

Processes in this RAS model correspond to "mission" trips executed by vehicles $v_i \in V$. They are formally defined by triplets: $< n_s, n_d, d_s >$, where

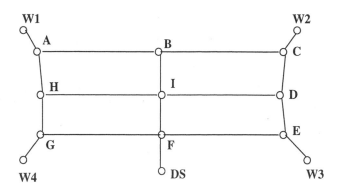

Figure 4.8.   The guidepath graph for the AGV system in Figure 1.3

$n_s, n_d \in N_W$ define the *"source"* and the *"destination"* nodes, respectively. The trips are initiated either at the docking station d_s, or by re-assigning and re-directing an *idle* vehicle performing the last part of its trip, i.e., returning from node n_d to node d_s. Obviously, at any point in time, the number of processes (vehicles) in the system is limited by $|V|$, the number of vehicles available in the system.

For logical analysis purposes, vehicles are assumed to always reside in one of the guidepath edges $e \in E$. Specifically, even though, in the physical operation of the system, a vehicle can find itself in one of the system intersections, this situation is considered to be *transient*, corresponding to transitions from link to link in the system guidepath network; vehicles cannot stay in an intersection while waiting for the acquisition of the next link. In other words, intersections are only facilitators of the motion from link to link, and they do not constitute resources of the underlying RAS model. The only control needed for the allocation of the intersection zones, is the localized prioritization of the vehicles waiting to cross them, so that collisions in the intersections are avoided.

Finally, given the above characterization of the constituent elements of the AGV RAS, a natural definition of its *size* is $|\Phi_{AGV}| \equiv |N| + |E| + |V|$.

FSA-based modelling of the RAS behavior.   A formal characterization of the AGV RAS *state* that is sufficient for the logical analysis of its behavior, is as follows:

DEFINITION 21 *The state s of the AGV RAS is defined by the status of each of the $|V|$ vehicles circulating in the system. Specifically, any vehicle $v_i \in V$ can be either* PARKED *at the docking station d_s, or* TRAVELLING, *on a "mission" trip. In the latter case, the vehicle status is further specified by: (i) an* ordered list L_{v_i} containing the "milestone" nodes remaining to be visited, and (ii) the guidepath edge, $e_{v_i} \equiv (k, l)$, currently allocated to the vehicle. Moreover, edge (k, l) should be interpreted as a directed edge, with the direction defining the current vehicle motion on it.

From Definition 21, and the previous description of the AGV RAS elements, it follows that the system state space, S, is *finite*. Furthermore, *transitions* between these states correspond to the following events:

1. A parked vehicle is assigned to a new "mission" and allocated its first zone.

2. A vehicle waiting on edge (i, j) is allocated a neighboring edge (j, k). In case that edge (j, k) is a link leading to a workstation terminal node, it is automatically assumed that the vehicle has veered its direction to (k, j). If $k \equiv d_s$, the vehicle status is switched to PARKED.

3. An idle vehicle on its way to the docking station d_s, is re-assigned and redirected to a new "mission" trip.

The RAS state space armed with the state transition function described above, defines a *Finite State Automaton (FSA)* (Cassandras and Lafortune, 1999; Hopcroft and Ullman, 1979). The *initial* and *final* states of this automaton correspond to state s_0, in which the system is at rest, i.e., all vehicles are parked at d_s. Hence, the *language* accepted by this automaton corresponds to sets of completed transport tasks.

AGV deadlock. As it was demonstrated in Example 2, the completion of some of these transport tasks might not be possible, under uncontrolled operation, due to the occurrence of deadlocks (cf. Fig. 1.3). The detailed characterization of the AGV deadlock and the development of the necessary nonblocking supervisors for its avoidance depends on whether the system operates under a *static* or a *dynamic* routing policy.[4] AGV RAS operated under a *static* routing scheme, are equivalent to LIN-SU-RAS. To see this, notice that the a priori specification of the detailed routes that should be followed when travelling between any pair of nodes of the guidepath graph, implies that every vehicle mission trip can be expanded to an exact sequence of guidepath edges that need to be allocated to the vehicle for the completion of its mission. Hence, all the theory developed in the previous parts of this book regarding the characterization and the avoidance of deadlock in LIN-SU-RAS applies immediately to the AGV RAS operated

---

[4]We remind the reader that under *static* routing, vehicles are travelling towards their milestone nodes on predetermined guidepath edge sequences, typically selected on the basis of some "shortest path" criterion, while under *dynamic* routing, the next guidepath edge to be allocated to a vehicle having completed the traversal of its current edge, is determined on-line, taking into consideration the prevailing traffic conditions in the network.

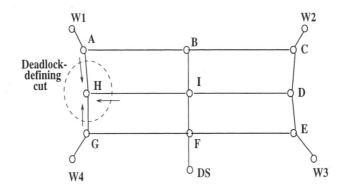

Figure 4.9. The cut of the guidepath graph corresponding to the deadlock of Figure 1.3

under static routing. Next, we address the most interesting case of AGV RAS operated under _dynamic_ routing.

A formal definition of the AGV deadlock in this subclass of AGV RAS, is as follows:

DEFINITION 22 _For AGV RAS operated under dynamic routing, deadlock corresponds to a RAS state in which there exists a_ cut _of the guidepath graph such that: (i) one of the defined subgraphs is allocated to capacity, and (ii) all the cut edges either are allocated to AGV's moving towards the fully allocated subgraph, or they are free edges linking some workstation node_ $n_i \in N_W$ _to the rest of the guidepath network, and they are not accessible by the vehicles allocated to neighboring edges._

The formal connection of the AGV deadlock concept, provided in Definition 22, to the blocking effects arising during the RAS operation, can be established through arguments similar to those provided in Section 2.2 of Chapter 2; we leave those technical details to the reader. It is interesting to notice that if one of the subgraphs defined by the cut considered in Definition 22 is a single node, condition (i) is trivially satisfied. This particular type of deadlock corresponds to deadlock arising at the intersections of the AGV guidepath network; for instance, Figure 4.9 indicates the cut corresponding to the AGV system deadlock depicted in Figure 1.3: it consists of edges (A, H), (I, H) and (G, H), with the sub-graph allocated to capacity being the single node H.

Complexity considerations and state unsafety. Detection of the AGV deadlock of Definition 22 can be achieved in polynomial time with respect to the size of the underlying RAS, through an algorithm similar to that presented in Figure 2.8, that iteratively identifies and eliminates from the current state repre-

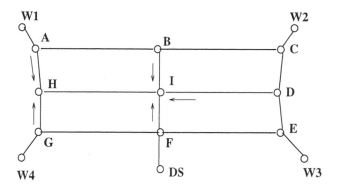

Figure 4.10. An unsafe state for the AGV RAS of Figure 1.3.

sentation any vehicle that can advance to a neighboring edge, while releasing its currently allocated edge to be used by the remaining vehicles. However, similar to the case of DIS-CON-RAS, to effectively avoid AGV deadlock states, one must avoid a broader class of *unsafe* states, i.e., AGV RAS states from which deadlock is unavoidable. An unsafe state for the AGV RAS of Figure 1.3 is depicted in Figure 4.10: The arrows along the edges of the depicted guidepath graph indicate the existence of a vehicle on the edge, moving along the indicated direction. Hence, for the depicted situation, the only feasible movement for these five vehicles is to edge (HI). Moreover, no matter which of the five vehicles is allocated this edge, the system will end up in deadlock.

Currently, it is unclear whether state safety in dynamically routed AGV RAS is polynomially decidable with respect to the AGV RAS size $|\Phi_{AGV}|$. Hence, it is not certain whether the *optimal* – i.e., the maximally permissive – SCP for this RAS is computationally tractable. Under these circumstances, the next section discusses a variation of Banker's algorithm which provides a polynomial, although suboptimal, SCP for this RAS.

3.2 The AGV Banker's Algorithm

The fundamental mechanism that defines Banker's algorithm implementation on the AGV RAS and establishes its correctness is identical to that underlying the algorithm implementation on the LIN-SU-RAS. The system operation is restricted in the subspace defined by the set of reachable ordered states, i.e., those states for which there exists a feasible event sequence that advances and completes active vehicle missions one at a time. Hence, the basic logic of the algorithm presented in Figure 4.4 carries over identical to the case of AGV RAS, when (i) the set of active processing stages $\mathcal{S}^a(s)$ is substituted by the set of active vehicle missions $V^a(s)$, and the resource set is understood to be the set of the guidepath edges E. However, one particular element of the algorithm that needs special consideration is step (2b), since, in the AGV RAS context, process routes are represented only implicitly, through the structure of the AGV guidepath network. The next proposition provides a detailed test for the implementation of this step.

PROPOSITION 3 *Consider a vehicle v in an AGV RAS state s, currently situated on edge $e \equiv (k, l)$ of the guidepath graph, and with its remaining "mission" trip being defined by the ordered list $L_u \equiv< n_1, (n_2 \vee NULL), (n_3 \vee NULL) >$. Then v can complete its mission trip, as requested by Step (2b) of the Banker's algorithm stated in Figure 4.4, iff state s satisfies the following two conditions:*

1 *There exists a path of free edges that connect node l to node n_1, or there exist a path of free edges that connect node k to node n_1 plus a circuit of free edges reachable from node l through a path of free edges.*

2 *There exist paths of free edges connecting each of the nodes n_2 and n_3 to one of the nodes l or k; if any of the nodes n_2 and n_3 is equal to $NULL$, the corresponding requirement is trivially satisfied.*

Proof: The truth of this proposition is an immediate consequence of the previously stated assumptions regarding the operation of the AGV RAS under dynamic vehicle routing. In particular, the satisfaction of Condition 1 is necessary and sufficient for achieving the first milestone in the vehicle mission, given the unidirectional motion of the vehicle on its currently allocated guidepath link. Assuming that Condition 1 has been satisfied, Condition 2 then becomes necessary and sufficient for the vehicle to access milestone nodes n_2 and n_3. The sufficiency of Condition 2 for accessibility of milestone nodes n_2 and n_3 is evident from the facts that (i) node n_1 is connected to at least one of the nodes k and l, and (ii) the freed edge (k, l) can be reused by the vehicle in any direction on its way to any of the two milestone nodes n_2 and n_3. To see the necessity of Condition 2, notice that both nodes n_2 and n_3 must be connected to node n_1. If the path from node n_1 to node n_2 does not engage edge (k, l) and Condition 1 was satisfied through its first part, then node l must be connected to node n_2 through a set of free edges in state s. If the path from node n_1 to node n_2 does not engage edge (k, l) and Condition 1 was satisfied through its second part, then node k must be connected to node n_2 through a set of free edges in state s. The remaining cases are similarly analyzed. ◇

Testing Conditions 1 and 2 of Proposition 3 on the AGV guidepath graph can be done efficiently through some *labelling* procedures. More specifically, the connectivity of node $i \in \{k, l\}$ to node n_j, $j = 1, 2, 3$, can be efficiently checked through a labelling scheme that initially labels node i and tries to construct a path to node n_j, by successively picking already labelled nodes, and using the edges emanating from them in order to "reach out" and label additional unlabelled nodes; the procedure accepts when node n_j is labelled, and it rejects when no further nodes can be labelled while node n_j has not been reached. Similarly, the existence of a circuit containing node $i \in N$ can be checked through a labelling procedure that utilizes a different label for the nodes reached from node i through each distinct edge emanating from it. The procedure terminates with a positive response if some node $j \not\equiv i$ is double-labelled. Further details regarding such labelling algorithms can be found in ((Ahuja et al., 1993), Sect. 3.4). In fact, one can develop an even more efficient procedure for testing the connectivity of the various nodes in the guidepath network through paths of free edges, by noticing that this property essentially establishes an *equivalence relationship* $\mathcal{E}$ on the node set N of the guidepath graph, with the equivalence classes C_i corresponding to the subsets of nodes connected to each other through currently free edges. The availability of a data structure pertinently encoding this equivalence relationship $\mathcal{E}$, at any point during the execution of Banker's algorithm, would render the testing of the connectivity of any pair of nodes i and j a task of complexity $O(1)$. Indeed, such a data structure can be easily developed and maintained during the execution of Banker's algorithm, by starting with a set of classes C_i where each class C_i contains the single node $i \in N$, and subsequently merging any pair of classes C_k, C_l, corresponding to a free or freed edge (k, l). We refer the reader to (Reveliotis, 2000b) for a more detailed discussion of this idea. Also, similar computational economies are possible with respect to testing the existence of circuits of free edges for the various nodes of the network.

Examples

We conclude this section by presenting two examples which demonstrate the notion of ordered AGV RAS states, and the detailed application of the AGV Banker's logic.

First, consider the situation depicted in Figure (4.11,a). In this case, vehicle v_1 has just unloaded a part at workstation W_3 and it is heading towards the docking station, vehicle v_2 is crossing the indicated zone of the guidepath graph transferring a part to workstation W_3, vehicle v_3 is heading towards workstation W_2 in order to pick a part that is to be transferred to workstation W_3, while vehicle v_4 is idling at the docking station. At the considered point, a new transfer request is posed to the AGV system controller, concerning the transfer of a part from workstation W_1 to workstation W_2. The controller considers allocating vehicle v_4 to this transfer, in which case, the vehicle should depart on the corresponding trip by traversing the guidepath zone between the docking station, and the workstation W_4. However, before commanding the initiation

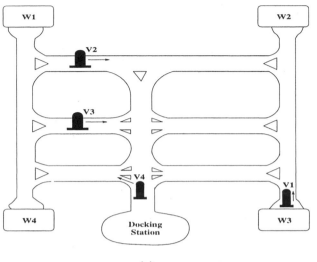

(a)

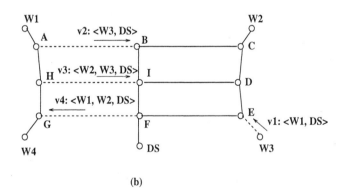

(b)

Figure 4.11. Example: Running the AGV Banker's algorithm on an ordered state

of this motion to the vehicle controller, the system controller must evaluate its safety with respect to the development of any potential deadlocks. To this end, the AGV system controller must determine whether the AGV RAS state that would result from the allocation of the aforestated zone to vehicle v_4 is *ordered*, i.e., whether it is accepted by the AGV Banker's algorithm, detailed in the previous paragraph.

The AGV RAS state s to be evaluated by the AGV Banker's algorithm is depicted in Figure (4.11,b). Under the depicted vehicle positioning, the guide-path subgraph induced by the system free links presents three inter-connected components, defining the node equivalence classes: $C_1 = \{W_1, A, H, G, W_4\}$, $C_2 = \{W_2, C, D, E, F, I, B, DS\}$, and $C_3 = \{W_3\}$. Application of the logic of Proposition 3 indicates that vehicle v_4 can complete its "mission" trip, visiting first node W_1, and then returning to node W_2 through the freed edge (G, F). Subsequently, vehicle v_1 can use the freed edge (G, F) to reach W_1 and the docking station, DS, in that order. It is easy to see that, by the time vehicle v_1 completes its trip, the guidepath subgraph induced by free and *freed* edges is fully connected, and therefore, the remaining vehicles v_2 and v_3 can complete their trip in any order. Hence, state s is ordered, and therefore, vehicle v_4 can proceed with the acquisition and traversal of zone FG.

Next consider the situation depicted in Figure (4.12,a). In this case, vehicle v_1 has just been loaded with a part destined to workstation W_1, while vehicle v_2 has just unloaded a part at workstation W_1. Workstation W_1 possesses also a part that needs to be transferred to workstation W_3, and the AGV system controller contemplates assigning vehicle v_2 to this transfer request. However, before this assignment takes place, the system controller must evaluate the admissibility of the resulting AGV RAS state, s', by the AGV Banker's algorithm. The formal representation of state s' is depicted in Figure (4.12,b). It should be clear that under the depicted node inter-connectivity of the system guidepath graph, no vehicle can complete its trip; hence, state s' is not ordered. Notice, however, that state s' is safe, since advancing any of the two vehicles to one of its neighboring edges results in an ordered state. From a theoretical standpoint, the existence of unordered safe states, demonstrated by this example, implies that, in the general case, the AGV Banker's algorithm will be a *suboptimal* solution to the AGV deadlock problem, i.e., some vehicle movements will be unnecessarily restricted by the SCP.

Concluding the discussion of this example, we notice that the apparent inefficiency of Banker's algorithm, indicated by the non-admissibility of the safe state s', can be effectively overcome at the "performance-control" level. Specifically, an efficient performance-oriented control policy should recognize that, under the described situation, the best move for vehicle v_2 would perhaps be to wait on edge (W_1, A), until vehicle v_1 is safely moved out of edge (W_3, E), and then undertake the new assignment. Hence, this example also demonstrates that *intentional idleness* can sometimes be the preferred option during the operation of the considered RAS.

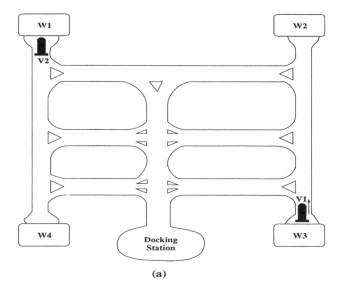

(a)

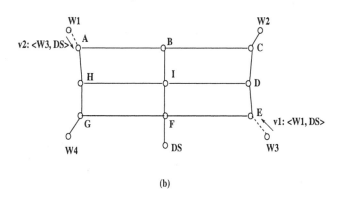

(b)

Figure 4.12. Example: Running the AGV Banker's algorithm on an unordered state

4.   Accommodating Operational Contingencies

In this section, we consider the issue of accommodating operational contingencies, like the loss of a resource unit or the arrival of an expedient process, in a LIN-SU-RAS supervised by the PK-SCP's introduced in this chapter. The presented results concretize in the considered operational context the basic ideas that were described at a more conceptual level in Section 5.2 of Chapter 2. Hence, in line with the discussion presented in that section, the material of this section is organized in two parts: The first part presents a *reactive* approach to the aforementioned problems, while the second part introduces a *proactive* framework.

4.1 The reactive approach

We remind the reader that the reactive approach essentially allows the problem to emerge, and subsequently, it seeks (i) to revise the applied SCP so that it accounts for the new developments, and (ii) to reconstruct the RAS state so that it is accepted by the new policy; this last step might involve the suspension of certain active processes and their (temporary) removal from the system. In the following, we discuss the execution of these two steps in the context of a LIN-SU-RAS supervised under a PK-SCP for the cases of

- the loss of a resource unit, and

- the appearance of an expedient process instance.

Accommodating the loss of a resource unit. For search-based PK-SCP's, this event does not necessitate the *explicit* revision of the policy definition. On the other hand, if RUN or RO SCP is applied, the loss of this resource unit must be reflected in the right-hand-side of the policy-defining constraints, since this vector is determined by the resource capacities. Once the new policy-defining condition, $\mathcal{H}'$, has been appropriately revised, the next step is the assessment of the admissibility of the current state s by $\Delta_{\mathcal{H}'}$, and, in case of a negative result, the removal of some processes from the system in order to bring the running RAS state into $S_r(\Delta_{\mathcal{H}'})$. This last decision can be rationalized by associating a *weight*, $w_j \geq 0$, with every active process instance j_j in s, to express a measure of the process criticality (or, equivalently, the "cost" of suspending this particular process). Then, if $\mathcal{I}$ denotes the entire set of active process instances in s, and $\mathcal{U}$ denotes the set of process instances to be removed from the system, a formal characterization of the considered problem is as follows:

$$\min_{\mathcal{U} \in 2^{\mathcal{I}}} c(\mathcal{U}) \equiv \sum_{j_j \in \mathcal{U}} w_j \tag{4.7}$$

s.t.

$$s(\mathcal{I} \backslash \mathcal{U}) \in S_r(\Delta_{\mathcal{H}'}) \tag{4.8}$$

In Equation 4.8, $s(\mathcal{I} \backslash \mathcal{U})$ denotes the RAS state resulting from state s after suspending the processes in $\mathcal{U}$. We shall refer to the problem defined by Equations 4.7 and 4.8 as the *"Optimal" Recovery* problem.

"Optimal" Recovery Algorithm
Input: LIN-SU-RAS Φ, a state s with a set of active process instances $\mathcal{I}$, a set of weights $w_j \geq 0$ for the process instances $j_j \in \mathcal{I}$, and an SCP $\Delta_{H'}$
Output: Minimum-weight process set, $\mathcal{U}^*$, to be unloaded so that $s(\mathcal{I} \backslash \mathcal{U}^*) \in S_r(\Delta_{\mathcal{H}'})$
Boolean Variable: DONE

1 Impose a total ordering on the set of process instances, $\mathcal{I}$, that is nondecreasing w.r.t. the job weights, w_j, i.e., define $h() : \mathcal{I} \rightarrow \{1, \ldots, |\mathcal{I}|\}$, s.t. $h(j_i) < h(j_j) \iff w_{j_i} \leq w_{j_j}$.

2 Set $\mathcal{A} := \{\{j_j\} : \forall j_j \in \mathcal{I}\}$; DONE := FALSE.

3 While (not DONE) do

   (a) $\mathcal{U}^* := \arg\min_{\mathcal{U} \in \mathcal{A}} \{c(\mathcal{U})\}$

   (b) If $(s(\mathcal{I} \backslash \mathcal{U}^*) \in S_r(\Delta_{\mathcal{H}'}))$, DONE := TRUE.

   (c) else $\mathcal{A} := (\mathcal{A} \backslash \{\mathcal{U}^*\}) \cup$
       $\{\mathcal{U}^* \cup \{j_j\} : \forall j_j \in \mathcal{I}$ with $h(j_j) > \max_{j_k \in \mathcal{U}^*} \{h(j_k)\}\}$

4 return $\mathcal{U}^*$

Figure 4.13.   An enumeration-based algorithm for solving the "Optimal" Recovery problem

Figure 4.13 presents an algorithm for the systematic solution of the "Optimal" Recovery problem. This algorithm solves the "Optimal" Recovery problem by enumerating the subsets of $\mathcal{I}$ in increasing order with respect to $c()$, until a subset $\mathcal{U}^*$ is found such that $s(\mathcal{I} \backslash \mathcal{U}^*) \in S_r(\Delta_{\mathcal{H}'})$. The next lemma and theorem formally establish this behavior and prove the algorithm correctness.

LEMMA 4 *During the execution of the "Optimal" Recovery algorithm, for every set $X \in 2^{\mathcal{I}}$ that has not already been discarded by the algorithm, there exists a set $\mathcal{U}_X \in \mathcal{A}$ such that $\mathcal{U}_X \subseteq X$.*

Proof: Given a set $X \in 2^{\mathcal{I}}$, order its elements in increasing order with respect to $h()$, and consider its subsets $X_{[i]}$, $i = 1, \ldots, |X|$, where subset $X_{[i]}$ contains the first i elements of X. Then, the initialization of set $\mathcal{A}$, in Step 2, combined with its updating formula in Step (3c), imply that either $X_{[i]} \in \mathcal{A}$ for some $i \in \{1, \ldots, |X|\}$, or X must have already been tested and discarded by the algorithm. $\diamond$

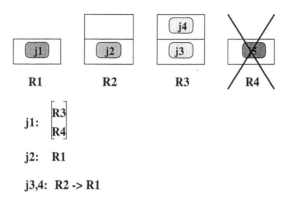

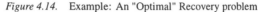

j1: $\begin{bmatrix} R3 \\ R4 \end{bmatrix}$

j2: R1

j3,4: R2 -> R1

Figure 4.14. Example: An "Optimal" Recovery problem

Table 4.2. Example: The RAS state

process instance	running stage	w_j
j_1	Ξ_{11}	10
j_2	Ξ_{22}	1
j_3	Ξ_{21}	5
j_4	Ξ_{21}	8
j_5	Ξ_{13}	-

THEOREM 4.5 *The process set $\mathcal{U}^*$ returned by the "Optimal" Recovery algorithm of Figure 4.13 is an optimal solution to the "Optimal" Recovery problem, defined by Equations 4.7 and 4.8.*

Proof: Theorem 4.5 is an immediate consequence of Lemma 4 and the non-negativity of the process weights w_j. ◇

The following example provides a concrete implementation of the above results, for the case of an SU-RAS logically controlled by the Banker's algorithm.[5]

Example. Consider the SU-RAS of Figure 4.14. This system consists of four resources, R_1, R_2, R_3, R_4, with $C_1 = C_4 = 1$ and $C_2 = C_3 = 2$. In

---

[5]Note that one of the process instances involved in this example possesses routing flexibility, and therefore, the considered RAS should be classified as a DIS-SU-RAS. However, it should be obvious to the reader that Banker's algorithm can be readily extended to DIS-SU-RAS through the employment of labelling procedures similar to those employed by the AGV Banker's algorithm.

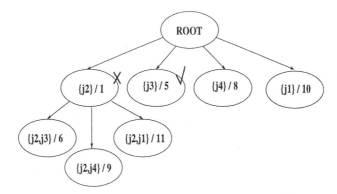

Figure 4.15.   Example: The search-tree developed by the "Optimal" Recovery algorithm

its current configuration, the system supports two process types: Process type Π_1 has a processing stage set $\mathcal{S}_1 = \{\Xi_{11}, \Xi_{12}, \Xi_{13}\}$ with resource allocation requests $A_{11} = (1,0,0,0)^T$, $A_{12} = (0,0,1,0)^T$ and $A_{13} = (0,0,0,1)^T$. Furthermore, the graph $\mathcal{G}_1$ for process Π_1 has node set $N_1 \equiv \mathcal{S}_1$ and edge set $E_1 \equiv \{\Xi_{11}, \Xi_{12}), (\Xi_{11}, \Xi_{13})\}$, i.e., stage Ξ_{11} can be succeeded by any of the other two stages, Ξ_{12} and Ξ_{13}. Process type Π_2 has $\mathcal{S}_2 = \{\Xi_{21}, \Xi_{22}, \Xi_{23}\}$ with resource allocation requests $A_{21} = (0,0,1,0)^T$, $A_{22} = (0,1,0,0)^T$ and $A_{23} = (1,0,0,0)^T$. The graph $\mathcal{G}_2$ for process Π_2 has node set $N_2 \equiv \mathcal{S}_2$ and edge set $E_2 \equiv \{\Xi_{21}, \Xi_{22}), (\Xi_{22}, \Xi_{23})\}$, i.e., stages Ξ_{21}, Ξ_{22} and Ξ_{23} are to be executed sequentially. The current loading of the system is described in Table 4.2. It is easy to see that the current RAS state is ordered, with a feasible process termination sequence being $< j_5, j_1, j_2, j_3, j_4 >$. At the considered time point, resource R_4 breaks down, and process j_5 is scrapped. The RAS state of the remaining subsystem $\{R_1, R_2, R_3\}$ is not ordered, under the new resource availability, and therefore, to bring the system back in the subspace of ordered states, we must suspend some of the running processes, $j_1, \ldots, j_4$.

Application of the "Optimal" Recovery algorithm on this state evolves as depicted in Figure 4.15. Process instances are ordered based on their associated weights, i.e., $h(j_1) = 4$, $h(j_2) = 1$, $h(j_3) = 2$, $h(j_4) = 3$. Set $\mathcal{A}$ is initialized to $\mathcal{A}_1 := \{\{j_i\}, \ i = 1, \ldots, 4\}$. During the first iteration, the algorithm considers the set $\mathcal{U}_1^* = \{j_2\}$, but it can be easily seen that the state resulting by removing process j_2 is still unordered. Therefore, $\mathcal{A}$ is updated to $\mathcal{A}_2 := \{\{j_i\}, \ i = 1, 3, 4\} \cup \{\{j_2, j_k\}, \ k = 1, 3, 4\}$, and $\mathcal{U}_2^* = \{j_3\}$, with $c(\mathcal{U}_2^*) = 5$. The state resulting by removing process j_3 is ordered, with a feasible process termination sequence being $< j_1, j_2, j_4 >$. Hence, an optimal set of processes that must be suspended in order to get an ordered RAS state, under the new resource availability, is $\mathcal{U}_2^* = \{j_3\}$. ◇

Since the "Optimal" Recovery algorithm of Figure 4.13 is based on an enumeration of the powerset of the set of active process instances $\mathcal{I}$, its worst-case complexity is *exponentially* related to the RAS size $|\Phi|$. It is expected, however, that the number of processes to be unloaded in order to bring the RAS state back to the policy-admissible region will be rather small, and therefore, the algorithm will be able to identify an optimal solution by considering a small subset of $2^{\mathcal{I}}$. It is also possible to *"thin"* the process set $\mathcal{I}$ that should be considered for suspension, based on the logic underlying the applied SCP. For instance, in the case of Banker's algorithm, $\mathcal{I}$ need not include processes residing on resources with positive slack capacity, and among those process instances executing the same processing stage on a fully allocated resource, only one with the smallest weight w_j needs to be included in $\mathcal{I}$. Based on these remarks, the entire subtree rooted at node $< \{j_2\}|1 >$ as well as the node $< \{j_4\}|8 >$ could have been eliminated *a priori* from the search tree of Figure 4.15.

The algorithm of Figure 4.13 is applicable even in the case that the underlying SCP is the *optimal* SCP discussed in Chapter 3. However, the application of the "Optimal" Recovery algorithm in systems controlled by the optimal SCP, must be preceded by the verification that the configuration resulting from the resource loss preserves the structure that enables the application of the optimal policy; otherwise, another PK-SCP should be adopted for the resulting configuration. We also notice that the algebraic nature of the RUN and RO SCP's enables an alternative formulation of the "Optimal" Recovery problem as a binary program. Employing such a formulation is especially useful for the case of RUN and RO implementations on SU-RAS where some active processes possess routing flexibility. More specifically, in DIS-SU-RAS, the admissibility of some state $s(\mathcal{I}/\mathcal{U})$ could be attained by a pertinent rerouting of the active processes, and this additional dimension of the problem complicates significantly the test implied by Step (3b) in the algorithm of Figure 4.13, under RUN or RO SCP; we refer the reader to (Reveliotis, 1999) for the detailed discussion of the relevant formulations and some efficient algorithms for their solution.

Accommodating a newly arrived expedient process instance. The problem of accommodating a newly arrived expedient process into the system can be formulated as an "Optimal" Recovery problem of the type defined by Equations 4.7 and 4.8, on a state where the expedient process has been allocated its first requested resource, and it is assigned a very large criticality weight w_j, e.g., larger than the total weight of all other processes. In this way, all the algorithms and observations developed in the previous paragraphs for the problem of accommodating the loss of a resource unit, carry over to this new contingency accommodation problem.

4.2   The proactive approach

Under the proactive approach, the system is controlled by a single SCP across the entire set of its configurations emerging from the occurrence of the various contingencies, i.e., resource outages and reinstatements. The applied policy anticipates the potential changes in the system configuration, and it manages the process flow through the RAS in a way that the feasibility of some minimal behavior is guaranteed in spite of these changes. A system that is able to meet these minimal specifications is said to *degrade gracefully* in the face of the occurring contingencies, and the control paradigm that seeks to establish this capability in failure-prone systems is known as *robust* control. The design of robust supervisory control policies has received very limited attention in the context of the logical control applications considered in this book. To the best of our knowledge, the only available results on this topic are coming from the work presented in (Lawley and Sulistyono, 2002; Sulistyono and Lawley, 2002). The rest of this section outlines these results.

The work of (Lawley and Sulistyono, 2002; Sulistyono and Lawley, 2002) addresses the problem of designing a robust supervisory control policy in the LIN-SU-RAS context, and under the assumption that *only one* resource R_f is prone to failure. More specifically, resources R_i in the RAS of (Lawley and Sulistyono, 2002; Sulistyono and Lawley, 2002) consist of some finite buffering capacity, C_i, and a single processor that supports the execution of the various processing stages taking place in R_i. Failure of resource R_f implies the temporary outage of the corresponding processor, until its repair and reinstatement. However, even under failure, the resource buffering capacity C_f is still available for storage of active processes that will request R_f for the execution of some of their processing stages. In this operational regime, the requirement that defines the aforementioned graceful degradation of the system is that while R_f is failed, the system can still support the execution of processes that do not require R_f for any of their processing stages. The mechanism employed towards meeting this requirement, is to maintain the system loaded in such a way, that all process instances heading towards the failure-prone resource can be accommodated, in case of R_f failure, to the buffers of those resources that support only processes engaging R_f in their execution, and therefore, are not necessary for the execution of the reduced subset of process types mentioned above. However, an additional requirement is that the processes which are stalled due to R_f failure, are accommodated to the aforementioned resources in such a manner that they can eventually proceed to completion upon the repair of R_f.

The above ideas and the resulting SCP can be formalized as follows: Let $\mathcal{S}(R_f)$ denote the RAS processing stages supported by resource R_f, and $\mathcal{S}^{FD}$ denote the set of *"failure-dependent"* stages, i.e., the stages in $\mathcal{S}(R_f)$, plus those processing stages that have a successor stage in $\mathcal{S}(R_f)$. Also, define the resource set $\mathcal{R}^{FD} \equiv \{R_i : \text{every processing stage } \Xi_{jk} \text{ supported by } R_i \text{ belongs}$

to $\mathcal{S}^{FD}$}; i.e., $\mathcal{R}^{FD}$ contains R_f and all those resources supporting only process instances that will eventually request R_f for the execution of some subsequent processing stage. These are the resources that are expected to function as buffers for the process instances that are stalled due to R_f failure. Finally, define the *neighborhood* NH_i of a resource $R_i \in \mathcal{R}^{FD}$ by $NH_i \equiv \{\Xi_{jk} : \Xi_{jk}$ is supported by R_i or it has a successor stage $\Xi_{j,k+\nu}$ supported by R_i and all intermediate processing stages $\Xi_{j,k+q}$, $q = 0, \ldots, \nu - 1$, are supported by resources in $\mathcal{R} \backslash \mathcal{R}^{FD}\}$. A process instance is said to be in some neighborhood NH_i *iff* its current processing stage belongs to NH_i. Notice that according to the above definitions, every processing stage $\Xi_{jk} \in \mathcal{S}^{FD}$, and its active process instances, belong to one and only one neighborhood NH_i.

In the light of the above definitions, the SCP of (Lawley and Sulistyono, 2002; Sulistyono and Lawley, 2002) is expressed by the following three constraints on the RAS state s:

Constraint 1 For every $R_i \in \mathcal{R}^{FD}$, the number of process instances in the resource neighborhood NH_i does not exceed its capacity C_i.

Constraint 2 For every resource pair $R_i, R_j \in \mathcal{R}^{FD}$, the total number of process instances in NH_i and NH_j should be *strictly* less than $C_i + C_j$.

Constraint 3 There is an ordering of the active processing stages $\Xi_{jk} \in \mathcal{S}^a(s)$ similar to that introduced in the discussion of Banker's algorithm (c.f., Definition 20), so that the active process instances for stages $\Xi_{jk} \in \mathcal{S}^a(s) \cap \mathcal{S}^{FD}$ can advance to the resource $R_i \in \mathcal{R}^{FD}$ of their corresponding neighborhood NH_i, while active process instances of the remaining processing stages, $\Xi_{jk} \in \mathcal{S}^a(s) \backslash \mathcal{S}^{FD}$, can advance to completion.

When combined, Constraints 1 and 3 above ensure that every process instance stalled in the system due to an outage of resource R_f can be accommodated in $\mathcal{R}^{FD}$, allowing the remaining part of the RAS, defined by $\mathcal{R} \backslash \mathcal{R}^{FD}$, to continue supporting the subset of process types not engaging resource R_f in their execution. Constraint 2 is crucial for ensuring that processes accommodated in $\mathcal{R}^{FD}$ can still proceed to completion upon the repair of resource R_f; the underlying mechanism establishing this capability is similar in nature to the mechanism establishing the correctness of RO SCP. A formal characterization and a proof of the policy correctness can be found in (Lawley and Sulistyono, 2002).

5. Historical and bibliographical notes

As it was pointed out in Section 5 of Chapter 1, the research on nonblocking supervision for LIN-SU-RAS was motivated in the early 90's from the need to establish deadlock-free buffer-space allocation in flexibly automated manufacturing systems. Two seminal contributions introducing the problem to the research community were those presented in (Viswanadham et al., 1990; Wysk et al., 1991). However, these works failed to address the complexity of the underlying problem and to provide a rigorous characterization of the policy correctness. The first works to introduce the concept of correct Polynomial Kernel-SCP – although not under this particular name – were those of (Banaszak and Roszkowska, 1988; Banaszak and Krogh, 1990). Subsequently, an extensive research activity thoroughly analyzed the problem of deadlock arising in LIN-SU-RAS, and provided effective solutions to it.

Regarding the results reported in this chapter, the RUN SCP, discussed in Section 1.1.0, was originally defined in (Gaarder, 1993). However, a rigorous characterization of the policy logic, including its generalization on the basis of the applied resource ordering, and a formal proof for its correctness by means of the DES logical control framework considered in this book, were first provided in (Reveliotis and Ferreira, 1996). The works of (Lawley et al., 1997b; Lawley et al., 1997c) further elaborated on the policy specification and undertook an experimental study of the policy efficiency with respect to the performance index of Equation 4.5. RO SCP was introduced and studied in (Lawley et al., 1998b). The original idea underlying Banker's algorithm was introduced by Dijkstra in (Dijkstra, 1965) and it was studied more extensively in (Habermann, 1969). The adaptation of the original Banker's algorithm to the LIN-SU-RAS context, that was discussed in Section 1.2.0, was introduced in (Lawley et al., 1998a). The work of (Lawley et al., 1998a) introduced also the idea discussed in Section 2.3, of combining PK-SCP's with limited lookahead search, in order to augment their flexibility; for a more extensive discussion on supervisory control based on partial search, the reader is referred to (Chung et al., 1992). The decomposition of the LIN-SU-RAS supervisory control problem on the basis of the connectivity of the system WPD was first introduced in (Fanti et al., 1998a). Results on the order selection problem for RUN and RO SCP's can be found in (Lawley et al., 1997c; Lawley et al., 1998b) and also in (Reveliotis, 1996); we shall return to this problem in Chapter 5, where RUN SCP is extended to the broader class of DIS-CON-RAS. Some additional notable efforts that define correct SCP's for LIN-SU-RAS are the works presented in (Hsieh and Chang, 1994; Ezpeleta et al., 1995; Fanti et al., 1997; Fanti et al., 1998b; Wu and Zhou, 2001). The policy introduced in (Hsieh and Chang, 1994) admits a RAS state s only if it can construct a process advancement sequence that successfully completes all active process instances. However, the search for such a termination sequence observes an additional process prioritization scheme that is supposed to be based on performance considerations and effectively constrains the system operation on a single path; due to this restriction, the policy presents polynomial complexity with respect to the underlying RAS size. The policies of (Ezpeleta et al., 1995; Wu and Zhou, 2001) and the most interesting of the policies developed in (Fanti et al., 1997; Fanti et al., 1998b) seek to explicitly prevent the formation of every single deadlock that can occur in the system, and as a result, they have a worst-case complexity that is non-polynomial with respect to the RAS size.

The problem of establishing nonblocking operation in the AGV RAS introduced in Section 3.1, is known as *"conflict-free routing"* in the relevant literature. As it was pointed out in the discussion of the relevant example in Chapter 1, currently it is primarily addressed through a simple-minded deadlock prevention scheme, that decomposes the guidepath graph in a number of unidirectional loops, interfaced through a number of exchange buffers. We refer the reader to (Ganesharajah et al., 1998) for a comprehensive discussion of the relevant issues and the prevailing traffic control practices. The AGV RAS characterization and the Banker's algorithm presented in Section 3 appeared originally in (Reveliotis, 2000b). Another set of very interesting results on the AGV supervisory control problem, which consider, however, a modified version of the AGV RAS presented in Section 3.1, in that vehicles never exit the system, can be found in (Roszkowska, 2003; Roszkowska, 2004). Finally, the work of (Park et al., 2002) addresses the integration of the various results developed in this chapter towards the development of a hierarchical event-driven control architecture appropriate for contemporary flexibly automated production systems, organized according to a cellular layout.

Contingency accommodation on sequential RAS has received very limited attention in the literature. The material on reactive methods for nonblocking supervisory control of SU-RAS, that was presented in Section 4.1, originally appeared in (Reveliotis, 1999), while the material of Section 4.2, on proactive methods, was taken from (Lawley and Sulistyono, 2002; Sulistyono and Lawley, 2002). Some additional interesting results regarding the existence of robust / fault-tolerant SCP's for some DES classes more general than the RAS class considered herein, can be found in (Park and Lim, 1999).

Chapter 5

LOGICAL CONTROL OF RAS WITH COMPLEX PROCESS FLOWS

This chapter addresses the problem of establishing nonblocking operation for the more complex classes of the RAS taxonomy introduced in Chapter 1, including the most complex class of CPX-CON-RAS. The analytical characterizations and the conceptual insights developed in Chapters 2-4 with respect to the logical control of the simpler class of DIS-SU-RAS, provide valuable stepping stones for tackling the more complex problem version addressed in this chapter. In fact, some supervisory control policies that will be eventually utilized for establishing the deadlock-free operation of the broader RAS classes considered in this chapter, are generalizations of the RUN SCP and the Banker's algorithm introduced in Chapter 4 for the LIN-SU-RAS. Yet, the complexity of the behavior generated by the RAS classes considered in this chapter necessitates the introduction of a more powerful modelling framework for representing the underlying RAS dynamics, and the systematic redefinition of the undertaken logical control problem in this new modelling paradigm. The particular modelling framework adopted for this purpose is that of *Petri Nets (PN)* (Murata, 1989). Beyond providing the necessary representational specificity, recasting of the RAS logical control problem in the PN framework also offers novel analytical insights and additional computational tools that can significantly expand our methodological capability towards the synthesis of SCP's appropriate for the entire spectrum of the RAS taxonomy considered in this book. In fact, an additional intention of this chapter is to demonstrate the gains that can be achieved by the synergistic utilization of more than one representational frameworks for the RAS logical control problem.

From an organizational standpoint, the chapter starts with a brief introduction to the PN modelling framework, presenting all the concepts and results that are necessary for the later developments. The PN modelling framework is subsequently employed for the formal characterization of the structure and the behavior of the considered RAS classes with respect to deadlock. A sufficiency test for asserting the absence of deadlock in the operation of any given RAS configuration is developed. The second part of the chapter undertakes the problem of establishing nonblocking operation for the considered RAS classes. Two main methodological approaches are pursued: The first approach seeks to synthesize algebraic SCP's appropriate for the considered RAS classes while using the deadlock-freedom sufficiency test mentioned above, in order to assess the correctness of any such tentative policy. The second approach reduces the control synthesis problem to a corresponding problem formulated in the context of DIS-CON-RAS, and it employs a generalization of RUN SCP or Banker's algorithm for the solution of the reduced problem. The last part of the section extends the presented results to RAS classes containing Type-1 and/or Type-2 uncontrollability (c.f. Section 5.1 in Chapter 2 for the definition of these concepts).

1.   Petri net-based modelling of the RAS behavior

This first part of this section provides an overview of the Petri net (PN) related concepts that are necessary for the formal modelling of the considered RAS classes and for the analysis of their deadlock-related properties. The detailed characterization of the PN structure modelling the considered resource allocation environments is provided in the second part of the section.

1.1   Petri nets: Basic Concepts and Definitions

A formal definition of the Petri net model is as follows:

DEFINITION 23 *(Murata, 1989) A* Petri net (PN) *is defined by a quadruple* $\mathcal{N}$ $= (P, T, W, M_0)$, *where*

- P *is the set of* places,

- T *is the set of* transitions,

- $W : (P \times T) \cup (T \times P) \rightarrow Z_0^+$ *is the* flow relation, *and*

- $M_0 : P \rightarrow Z_0^+$ *is the net* initial marking, *assigning to each place* $p \in P$, $M_0(p)$ *tokens.*

The first three items in Definition 23 essentially define a *weighted bipartite digraph* representing the system *structure* that governs its underlying dynamics. The last item defines the system *initial state*. A conventional graphical representation of the net structure and its marking depicts nodes corresponding to places by empty circles, nodes corresponding to transitions by bars, and the tokens located at the various places by small filled circles. The flow relation W is depicted by directed edges that link every nodal pair for which the corresponding W-value is non-zero. These edges point from the first node of the corresponding pair to the second, and they are also labelled – or, *"weighed"* – by the corresponding W-value. By convention, absence of a label for any edge implies that the corresponding W-value is equal to unity.

PN structure-related concepts and properties. Given a transition $t \in T$, the set of places p for which $(p, t) > 0$ (resp., $(t, p) > 0$) is known as the set of *input* (resp., *output*) places of t. Similarly, given a place $p \in P$, the set of transitions t for which $(t, p) > 0$ (resp., $(p, t) > 0$) is known as the set of *input* (resp., *output*) transitions of p. It is customary in the PN literature to denote the set of input (resp., output) transitions of a place p by ${}^\bullet p$ (resp., $p^\bullet$). Similarly, the set of input (resp., output) places of a transition t is denoted by ${}^\bullet t$ (resp., $t^\bullet$). This notation is also generalized to any set of places or transitions, X, e.g. ${}^\bullet X = \bigcup_{x \in X} {}^\bullet x$.

The ordered set $X = < x_1 \ldots x_n > \in (P \cup T)^*$ is a *path*, if and only if (*iff*) $x_{i+1} \in x_i^\bullet, i = 1, \ldots, n - 1$. Furthermore, a path X is characterized as a *circuit iff* $x_1 \equiv x_n$.

A PN with a flow relation W mapping onto $\{0, 1\}$ is said to be *ordinary*. If only the restriction of W to $(P \times T)$ maps on $\{0, 1\}$, the PN is said to be *PT-ordinary*. An ordinary PN such that (s.t.) $\forall t \in T, |t^\bullet| = |{}^\bullet t| = 1$, is characterized as a *state machine*, while an ordinary PN s.t. $\forall p \in P, |p^\bullet| = |{}^\bullet p| = 1$, is characterized as a *marked graph*.

A PN is said to be *pure* if $\forall (x, y) \in (P \times T) \cup (T \times P), W(x, y) > 0 \Rightarrow W(y, x) = 0$. The flow relation of pure PN's can be represented by the *flow matrix* $\Theta = \Theta^+ - \Theta^-$ where $\Theta^+(p, t) = W(t, p)$ and $\Theta^-(p, t) = W(p, t)$.

PN dynamics-related concepts and properties. In the PN modelling framework, the system state is represented by the net *marking* M, i.e., a function from P to Z_0^+ that assigns a *token* content to the various net places. The net marking M is initialized to marking M_0, introduced in Definition 23, and it subsequently evolves through a set of rules summarized in the concept of *transition firing*. A concise characterization of this concept has as follows: Given a marking M, a transition t is *enabled iff* for every place $p \in {}^\bullet t$, $M(p) \geq W(p, t)$, and this is denoted by $M[t\rangle$. $t \in T$ is said to be *disabled* by a place $p \in {}^\bullet t$ at M *iff* $M(p) < W(p, t)$. Furthermore, a place $p \in P$ for which there exists $t \in p^\bullet$ s.t. $M(p) < W(p, t)$ is said to be a *disabling* place at M. Given a marking M, a transition t can be *fired* only if it is enabled in M, and firing such an enabled transition t results in a new marking M', which is obtained from M by removing $W(p, t)$ tokens from each place $p \in {}^\bullet t$, and placing $W(t, p')$ tokens in each place $p' \in t^\bullet$. For pure PN's, the marking evolution incurred by the firing of a transition t can be concisely expressed by the *state equation*:

$$M' = M + \Theta \cdot \mathbf{1}_t \qquad (5.1)$$

where $\mathbf{1}_t$ denotes the unit vector of dimensionality $|T|$ and with the unit element located at the component corresponding to transition t.

The set of markings reachable from the initial marking M_0 through any *fireable* sequence of transitions is denoted by $R(\mathcal{N}, M_0)$ and it is referred to as the net *reachability space*. In the case of pure PN's, a necessary condition for $M \in R(\mathcal{N}, M_0)$ is that the following system of equations is feasible in z:

$$M = M_0 + \Theta z \qquad (5.2)$$

$$M \geq 0, \; z \in Z_0^+ \qquad (5.3)$$

A PN $\mathcal{N} = (P, T, W, M_0)$ is said to be *bounded iff* all markings $M \in R(\mathcal{N}, M_0)$ are bounded. $\mathcal{N}$ is said to be *structurally bounded iff* it is bounded for any initial marking M_0. $\mathcal{N}$ is said to be *reversible iff* $M_0 \in R(\mathcal{N}, M)$, for all $M \in R(\mathcal{N}, M_0)$. A transition $t \in T$ is said to be *live iff* for all $M \in R(\mathcal{N}, M_0)$, there exists $M' \in R(\mathcal{N}, M)$ s.t. $M'[t\rangle$; non-live transitions are said to be *dead* at those markings $M \in R(\mathcal{N}, M_0)$ for which there is no $M' \in R(\mathcal{N}, M)$ s.t. $M'[t\rangle$. PN $\mathcal{N}$ is *quasi-live iff* for all $t \in T$, there exists $M \in R(\mathcal{N}, M_0)$ s.t. $M[t\rangle$; it is *weakly live iff* for all $M \in R(\mathcal{N}, M_0)$, there exists $t \in T$ s.t. $M[t\rangle$; and it is *live iff* for all $t \in T$, t is live. A marking $M \in R(\mathcal{N}, M_0)$ is a (total) *deadlock iff* every $t \in T$ is dead at M.[1]

Siphons and their role in the interpretation of the PN deadlock.   Of particular interest for the liveness analysis of the PN's to be considered in this book is a structural element known as *siphon*, i.e., a set of places $S \subseteq P$ such that $^\bullet S \subseteq S^\bullet$. A siphon S is *minimal iff* there exists no other siphon S' s.t. $S' \subset S$. A siphon S is said to be *empty* at marking M *iff* $M(S) \equiv \sum_{p \in S} M(p) = 0$. S is said to be *deadly marked* at marking M, *iff* every transition $t \in {}^\bullet S$ is disabled by some place $p \in S$. Clearly, empty siphons are deadly marked siphons. It is easy to see that, if S is a deadly marked siphon at some marking M, then (i) $\forall t \in {}^\bullet S$, t is a dead transition in M, and (ii) $\forall M' \in R(\mathcal{N}, M)$, S is deadly marked. The next theorem connects total deadlocks arising in PN's to deadly marked siphons.[2]

THEOREM 5.1 *Given a deadlock marking M of a PN $\mathcal{N} = (P, T, W, M_0)$, the set of disabling places $S \subseteq P$ in M constitutes a deadly marked siphon.*

---

[1] Notice that the concept of *deadlock* in the PN framework is different from the term usage in the RAS context. We shall further elaborate on this difference in a following section.

[2] This theorem constitutes a generalization of a well-established relationship between total deadlocks and empty siphons in ordinary PN's (Desel and Esparza, 1995).

Proof: We need to show that (i) $^\bullet S \subseteq S^\bullet$ and (ii) for all $t \in {}^\bullet S$, t is disabled by some place $p \in S$. To see the validity of the first statement, just notice that, since S contains all the disabling places and M is a deadlock marking, $S^\bullet = T$. The validity of the second statement results immediately from the definition of S and the fact that every transition t is dead in M. $\diamond$

PN semiflows. PN semiflows provide an analytical characterization of various concepts of *invariance* underlying the net dynamics. Generally, there are two types, p and t-semiflows, with a *p-semiflow* formally defined as a $|P|$-dimensional vector y satisfying $y^T \Theta = 0$ and $y \geq 0$, and a *t-semiflow* formally defined as a $|T|$-dimensional vector x satisfying $\Theta x = 0$ and $x \geq 0$. In the light of Equation 5.2, the invariance property expressed by a p-semiflow y is that $y^T M = y^T M_0$, for all $M \in R(\mathcal{N}, M_0)$. Similarly, Equation 5.2 implies that for any t-semiflow x, $M = M_0 + \Theta x = M_0$.

Given a p-semiflow y (resp., t-semiflow x) its *support* is defined as $\|y\| = \{p \in P \mid y(p) > 0\}$ (resp., $\|x\| = \{t \in T \mid x(t) > 0\}$). A p-semiflow y (resp., t-semiflow x) is said to be *minimal iff* there is no p-semiflow y' (resp., t-semiflow x') s.t. $\|y'\| \subset \|y\|$ (resp., $\|x'\| \subset \|x\|$).

PN merging. We conclude our general discussion on the PN concepts and properties to be employed in the subsequent parts of this work, by introducing a merging operation of two PN's: Given two PN's $\mathcal{N}_1 = (P_1, T_1, W_1, M_{01})$ and $\mathcal{N}_2 = (P_2, T_2, W_2, M_{02})$ with $T_1 \cap T_2 = \emptyset$ and $P_1 \cap P_2 = Q \neq \emptyset$ s.t. for all $p \in Q$, $M_{01}(p) = M_{02}(p)$, the PN $\mathcal{N}$ resulting from the *merging* of the nets $\mathcal{N}_1$ and $\mathcal{N}_2$ *through the place set* Q, is defined by $\mathcal{N} = (P_1 \cup P_2, T_1 \cup T_2, W_1 \cup W_2, M_0)$ with $M_0(p) = M_{01}(p)$, $\forall p \in P_1 \backslash P_2$; $M_0(p) = M_{02}(p)$, $\forall p \in P_2 \backslash P_1$; $M_0(p) = M_{01}(p) = M_{02}(p)$, $\forall p \in P_1 \cap P_2$.

1.2 PN-based modelling of the RAS behavior

This section provides a PN-based representation for the entire class of RAS encompassed by Definition 1. The development of this model proceeds in three stages: (i) first a PN model for capturing the execution logic of any single process instance is developed; (ii) subsequently, this model is augmented with resource places in order to represent the dynamics of the associated resource allocation; and finally (iii) the complete RAS model is obtained by merging the various subnets developed in step (ii) through their common resource places.

PN-based modelling of the RAS process types. In the PN modelling framework, the process type $\Pi_j = < \mathcal{S}_j, \mathcal{G}_j >$, introduced in item 3 of Definition 1, will be represented by the concept of the *process subnet*, formally defined as follows:

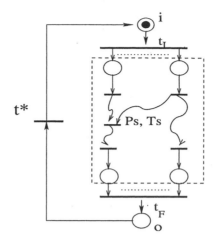

Figure 5.1.   The process net structure of Definition 24

DEFINITION 24 *A process (sub-)net is an ordinary Petri net* $\mathcal{N}_P = (P, T, W, M_0)$ *such that:*

  i $P = P_S \cup \{i,\ o\}$ *with* $P_S \neq \emptyset;$

 ii $T = T_S \cup \{t_I,\ t_F,\ t^*\};$

iii $i^{\bullet} = \{t_I\};\ ^{\bullet}i = \{t^*\};$

 iv $o^{\bullet} = \{t^*\};\ ^{\bullet}o = \{t_F\};$

  v $t_I^{\bullet} \subseteq P_S;\ ^{\bullet}t_I = \{i\};$

 vi $t_F^{\bullet} = \{o\};\ ^{\bullet}t_F \subseteq P_S;$

vii $(t^*)^{\bullet} = \{i\};\ ^{\bullet}(t^*) = \{o\};$

viii *the underlying digraph is* strongly connected*;*

 ix $M_0(i) > 0\ \wedge\ M_0(p) = 0,\ \forall p \in P\backslash\{i\};$

  x $\forall M \in R(\mathcal{N}_P, M_0),\ M(i) + M(o) = M_0(i) \Longrightarrow M(p) = 0,\ \forall p \in P_S.$

The PN-based process representation introduced by Definition 24 is depicted in Figure 5.1. Process instances waiting to initiate processing are represented by tokens in place i, while the initiation of a process instance is modelled by the firing of transition t_I. Similarly, tokens in place o represent completed process instances, while the event of a process completion is modelled by the firing of transition t_F. Transition t^* allows the token re-circulation – i.e., the token transfer from place o to place i – in order to model *repetitive* process execution. Finally, the part of the net between transitions t_I and t_F, that involves the process places P_S, models the sequential logic defining the considered process type. In particular, places $p \in P_S$ correspond to the various processing stages $\Xi_{jk} \in \mathcal{S}_j$, while the net connectivity among these places concretizes component $\mathcal{G}_j$ of Π_j (c.f. item (3b) of Definition 1). As it can be seen in Definition 24, this part of the process subnet can be quite arbitrary. However, in order to capture the requirements posed by Conditions 1 to 3 of Section 2.1 in Chapter 1, we further qualify the considered process subnets through the following three assumptions:

ASSUMPTION 1 *The process subnets considered in this work are assumed to be* acyclic, *i.e., the removal of transition t^* from them renders them acyclic digraphs.*

ASSUMPTION 2 *The process subnets considered in this work are assumed to be* quasi-live *for $M_0(i) = 1$.*

ASSUMPTION 3 *The process subnets considered in this work are assumed to be* strongly reversible, *i.e., their initial marking M_0 can be reached from any marking $M \in R(\mathcal{N}_P, M_0)$, through a firing sequence that does not contain transition t_I.*

Assumption 1 is introduced in order to satisfy Condition 1 of Section 2.1 in Chapter 1. Assumption 2 pertains to the satisfaction of Condition 2, by essentially stipulating that, in the considered process subnets, every transition models a meaningful event that can actually occur during the execution of some process instance, and therefore, it is not redundant. Assumption 3 pertains to the satisfaction of Condition 3, as it essentially stipulates that, at any point in time and under expedient resource allocation, all *active* process instances can advance to completion. Since the main focus of this work is on the analysis and control of the resource allocation function taking place in the considered environments, we forego the further study of the process subnets themselves, and an extensive investigation of the structural and behavioral properties implied by Definition 24 and Assumptions 1-3. The interested reader can find some relevant discussion and results in (Van der Aalst, 1996; Van der Aalst, 1997; Van der Aalst and Van Hee, 2002; Van Hee et al., 2003; Jeng et al., 2002).

PN-based modelling of the resource allocation function. The modelling of the resource allocation associated with the process stage Ξ_{jk} corresponding to any place $p \in P_S$, necessitates the augmentation of the process subnet $\mathcal{N}_P$, defined above, with a set of *resource* places $P_R = \{r_l,\ l = 1, \ldots, m\}$, of initial marking $M_0(r_l) = C_l,\ l = 1, \ldots, m$, and with the corresponding flow sub-matrix, Θ_{P_R}, expressing the allocation and de-allocation of the various

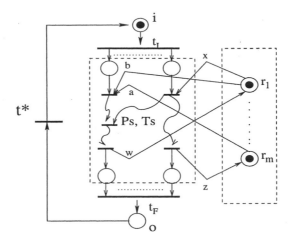

Figure 5.2.   The resource-augmented process net

resources to the process instances as they advance through their processing stages. The resulting net will be called the *resource-augmented process (sub-)net* and it will be denoted by $\overline{\mathcal{N}_P}$. Its basic structure is depicted in Figure 5.2. Notice that the interpretation of the role of transitions t^*, t_I and t_F provided in the previous section, implies that $(t^*)^\bullet \cap P_R = {}^\bullet(t^*) \cap P_R = (t_I)^\bullet \cap P_R = {}^\bullet(t_F) \cap P_R = \emptyset$. The *reusable* nature of the system resources is modelled by the following assumption regarding the resource-augmented process net $\overline{\mathcal{N}_P}$:

ASSUMPTION 4 *Let* $\overline{\mathcal{N}_P} = (P_S \cup \{i, o\} \cup P_R, T, W, M_0)$ *denote a resource-augmented process (sub-)net. Then,* $\forall l \in \{1, \ldots, |P_R|\}$, *there exists a p-semiflow* y_{r_l} *s.t.: (i)* $y_{r_l}(r_l) = 1$; *(ii)* $y_{r_l}(r_j) = 0$, $\forall j \neq l$; *(iii)* $y_{r_l}(i) = y_{r_l}(o) = 0$; *(iv)* $\forall p \in P_S$, $y_{r_l}(p) = $ *number of units from resource* R_l *required for the execution of the processing stage modelled by place* p.

While the p-semiflows introduced by Assumption 4 characterize the resource allocation taking place at each process stage and the conservative nature of the system resources, they do not provide any information regarding the adequacy of the available resource units for supporting the execution of the various processing stages, under the sequencing constraints implied by the process-defining logic. This additional concern underlying the correct definition of the various RAS process-types is captured by extending the requirement for *quasi-liveness* of the process net $\mathcal{N}_P$, introduced by Assumption 2, to the resource-augmented process net $\overline{\mathcal{N}_P}$:

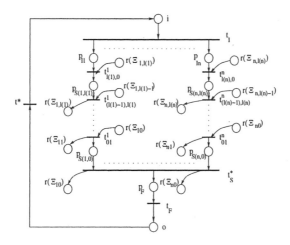

Figure 5.3. Reducing LIN-SU-RAS safety to the quasi-liveness of COR-RAS: The induced resource-augmented process net (resource places are repeated for clarity)

ASSUMPTION 5 *The resource-augmented process subnets considered in this work are assumed to be* quasi-live *for $M_0(i) = 1$ and $M_0(r_l) = C_l$, $\forall l \in \{1, \ldots, |P_R|\}$.*

Assessing the quasi-liveness of any given resource-augmented process net corresponding to a LIN or DIS-RAS, is a task of polynomial complexity with respect to the underlying RAS size. All that needs to be done is to ensure that for every resource R_i, $C_i \geq \max_{j,k}\{A_{jk}(i)\}$, or equivalently in the PN formalism, that for all $r_l \in P_R$ and for all $p \in P_{\mathcal{E}}$, $M_0(r_l) \geq y_{r_l}(p)$, where y_{r_l} are the p-semiflows introduced in Assumption 4. However, the problem of assessing the quasi-liveness of a resource-augmented process net belonging to the COR-RAS class is NP-hard (Garey and Johnson, 1979). An initial proof of this result was presented in (Roszkowska and Wojcik, 1993). Here we present an alternative proof that also reveals the main source for this increased complexity.

THEOREM 5.2 *The problem of deciding the quasi-liveness of a resource- augmented process net, $\overline{\mathcal{N}_P}$, belonging to the class of COR-RAS, is NP-hard.*

Proof: We establish this result through polynomial reduction of the LIN-SU-RAS safety problem, that was shown to be NP-complete in Section 3 of Chapter 2. Hence, consider a LIN-SU-RAS Φ with a resource set $\mathcal{R} = \{R_1, \ldots, R_m\}$, each resource available at capacity $C_l \geq 1$, and a state $s \neq s_0$ containing a set of active process instances $\mathcal{J} = \{J_1, \ldots, J_n\}$. Each process instance J_j, $j = 1, \ldots, n$, is executing processing stage Ξ_{j0}, and it is also associated with

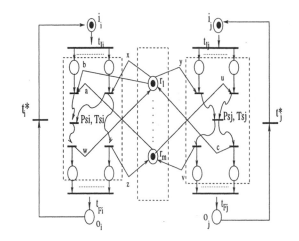

Figure 5.4.   The process-resource net structure considered in this work

a remaining processing stage sequence $< \Xi_{j1}, \ldots, \Xi_{j,l(j)} >$. Finally, each stage Ξ_{jk}, $j = 1, \ldots, n, k = 0, \ldots, l(j)$, requires for its execution the allocation of a single unit of a single resource $R(\Xi_{jk})$. The problem of determining the safety of state s is reduced polynomially to the problem of determining the quasi-liveness of a resource-augmented process net of the COR-RAS type, through the construction depicted in Figure 5.3.

The net of Figure 5.3 is a resource-augmented process net with its resource places being in one-to-one correspondence with the resources of the original LIN-SU-RAS safety problem. Furthermore, the subnet defined by nodes in $P_S \cup T_S$ consists of a set of $n(= |\mathcal{J}|)$ paths, with each path χ_j, $j = 1, \ldots, n$, and its associated resource allocation obtained by reversing the remaining processing stage sequence corresponding to active process instance J_j. In addition, all these paths, χ_j, concur to transition t_S^*. The firing of transition t_S^* releases all the allocated resources and enables the terminating transition t_F. The net initial marking is defined by $M_0(i) = 1$; $M_0(r_l) = C_l$, $l = l, \ldots, m$; $M_0(p) = 0$, otherwise. We claim that the constructed net is quasi-live *iff* the original LIN-SU-RAS state s is safe.

To establish this claim, first notice that the quasi-liveness of the constructed net essentially reduces to the quasi-liveness of transition t_S^*. Furthermore, a transition sequence, σ, enabling t_S^* from M_0, can be obtained from any existing safe – i.e., terminating – sequence for the original LIN-SU-RAS safety problem, by reversing the occurrence of the corresponding process advancing events. Similarly, any transition sequence, σ, enabling t_S^* from M_0, when reversed, provides a safe sequence for the original LIN-SU-RAS safety problem. ◇

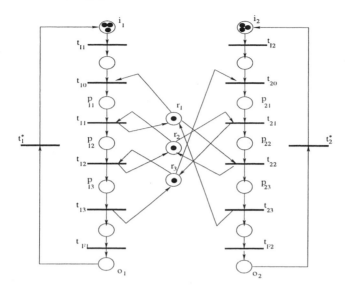

Figure 5.5. Example: The process-resource net modelling the LIN-SU-RAS of Section 1.3 in Chapter 1

The complete RAS model: Process-resource nets. The complete PN-based model, $\mathcal{N} = (P, T, W, M_0)$, of any given instance from the RAS class considered in this work is obtained by *merging* the resource-augmented process nets $\overline{\mathcal{N}_{P_j}} = (P_j, T_j, W_j, M_{0_j})$, $j = 1, \ldots, n$, modelling its constituent process types, through their common resource places. The resulting PN class is characterized as the class of *process-resource nets with acyclic, quasi-live and strongly reversible process subnets*, and its basic structure is depicted in Figure 5.4. Let $P = \bigcup_j P_j$, $P_S = \bigcup_j P_{S_j}$; $I = \bigcup_j \{i_j\}$; $O = \bigcup_j \{o_j\}$. Then, $P = P_S \cup I \cup O \cup P_R$. The re-usable nature of the resource allocation taking place in the entire process-resource net is characterized by a p-semiflow y_{r_l} for each resource type R_l, $l = 1, \ldots, m$, defined by: (i) $y_{r_l}(r_l) = 1$; (ii) $y_{r_l}(r_j) = 0$, $\forall j \neq l$; (iii) $y_{r_l}(i_j) = y_{r_l}(o_j) = 0$, $\forall j$; (iv) $\forall p \in P_S$, $y_{r_l}(p) = y_{r_l}^{(j*)}(p)$, where $\overline{\mathcal{N}_{P_{j*}}}$ denotes the resource-augmented process subnet containing place p, and $y_{r_l}^{(j*)}()$ denotes the corresponding p-semiflow for resource R_l. Furthermore, it is easy to see that Assumption 5, regarding the quasi-liveness of the constituent resource-augmented process subnets $\overline{\mathcal{N}_{P_j}}$, implies also the quasi-liveness of the entire process-resource net $\mathcal{N}$. Finally, in the PN modelling framework, the size of a RAS Φ, modelled by a net $\mathcal{N} = (P_S \cup I \cup O \cup P_R, T, W, M_0)$, is defined as $|\Phi| = |\mathcal{N}| \equiv |P_R| + |P_S| + \sum_{r \in P_R} M_0(r)$.

Example. Figure 5.5 depicts the process-resource net modelling the RAS considered in Section 1.3 of Chapter 1, in its initial state where the system is idle and empty of any processes. In the depicted net, we have set $M_0(i_j) = 3$ for $j = 1, 2$, so that these values do not constrain artificially the system loading. More generally, in the proposed PN-based RAS modelling, $M_0(i_j)$ must be set to a value that is an *upper bound* to the maximum number of process instances from process type Π_j that can be simultaneously loaded in the system. Such an upper bound will always exist due to the finiteness of the system resources. ◇

Characterizing the RAS taxonomy of Chapter 1 in the PN modelling framework. It is interesting, but also useful for the subsequent developments of this chapter, to identify the special structure that characterizes, in the PN modelling framework, the various classes of the RAS taxonomy presented in Chapter 1. The *Linear*, *Disjunctive* and *Coordinating* classes of that taxonomy, are respectively obtained by stipulating that the employed process nets are (i) *simple circuits* satisfying the requirements (i)-(ix) of Definition 24, (ii) *state machines* satisfying the requirements (i)-(ix) of Definition 24 and with every circuit containing transition t^*, and (iii) *marked graphs* satisfying the requirements (i)-(ix) of Definition 24 and with every circuit containing transition t^*. For all these three cases, it can be easily shown that the corresponding process nets also satisfy requirement (x) of Definition 24 as well as Assumptions 1-3.

The RAS classification on the basis of the structure of the resource allocation requests associated with the various processing stages, is modelled in the PN framework by imposing additional structure on the p-semiflows introduced by Assumption 4. More specifically, process-resource nets satisfying Assumption 4 without any additional conditions, correspond to the broadest class of *Conjunctive* RAS. *Single-Type* RAS are characterized by the fact that every place $p \in P_S$ belongs in the support of only one such semi-flow, while *Single-Unit* RAS further stipulate that $y_{r_l}(p) \in \{0, 1\}$ for all $p \in P_S$ and $r_l \in P_R$.

Modified markings of process-resource nets. Another concept related to the process-resource net, that is useful for the study of its deadlock-related properties undertaken in the next section, is that of the *modified marking*, formally defined as follows:

DEFINITION 25 *Given a process-resource net* $\mathcal{N} = (P_S \cup I \cup O \cup P_R, T, W, M_0)$ *and a marking* $M \in R(\mathcal{N}, M_0)$, *the* modified marking $\overline{M}$ *is defined by*

$$\overline{M}(p) = \begin{cases} M(p) & \text{if } p \notin I \cup O \\ 0 & \text{otherwise} \end{cases} \qquad (5.4)$$

Furthermore, the set of all modified markings induced by the net reachable markings is defined by $\overline{R(\mathcal{N}, M_0)} = \{\overline{M} \mid M \in R(\mathcal{N}, M_0)\}$

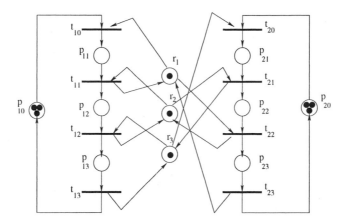

Figure 5.6. Example: A more compact PN-based representation for the LIN-SU-RAS of Section 1.3 in Chapter 1

The process idle place and concise process-resource net representations. Concluding this section, we notice that it is customary in the literature to aggregate the path $< i, t^*, o >$ of any process net to a single place p_0, that is called the process *idle place.* In that case, the set of places $I \cup O$ in the definition of the process-resource net $\mathcal{N}$ is substituted by the set P_0, containing all the process idle places. Clearly, all the definitions, assumptions and remarks regarding the basic net structure and properties translate directly to the modified net. In the following, both representations will be used interchangeably. More specifically, while the structure of the process net introduced in Definition 24 is useful for providing a clear canonical characterization of the underlying process behavior, in many cases it introduces redundant elements, increasing unnecessarily the number of the employed places and/or transitions. As a case in point, Figure 5.6 provides a more compact PN-based representation for the LIN-SU-RAS of Section 1.3, in Chapter 1, than that provided in Figure 5.5; the reader can convince herself that for the purposes of the logical analysis pursued in this work, the two representations of Figures 5.5 and 5.6 are equivalent. For the sake of conciseness, in the subsequent examples, we shall employ a compact model, whenever possible; the augmentation of the presented net structure with some additional places and/or transitions so that it matches the requirements of Definition 24 should be quite straightforward.

2. Analyzing the RAS deadlock in the PN-based modelling framework

It should be clear to the reader that in the modelling framework of process-resource nets with acyclic, quasi-live and strongly reversible process subnets, the requirement for nonblocking behavior is expressed by the requirement for *reversibility* of the RAS-modelling PN. However, one of the key results of this section is that, in the considered PN sub-class, liveness and reversibility are equivalent concepts. Therefore, the study of the problem of establishing RAS nonblocking behavior has primarily focused on the liveness of the RAS-modelling nets, and it has come to be known in the relevant PN literature as the problem of *liveness-enforcing supervision (LES)*.

This section provides a structural characterization for the liveness of process-resource nets with acyclic, quasi-live and strongly reversible process subnets, and in the process, it also establishes the aforementioned equivalence of liveness and reversibility for the considered PN sub-class. It is shown that the non-liveness of the considered process-resource nets can be interpreted through the emergence of a particular type of deadly marked siphon in the space of modified reachable markings. This finding subsequently allows the development of a sufficiency condition that can assert the nonblocking behavior of any given process-resource net with acyclic, quasi-live and strongly reversible process subnets. The derived condition takes the form of a mathematical programming formulation, and, as it will be shown in the next section, it is very useful for the correctness verification of tentative algebraic SCP's synthesized for any given configuration from the considered RAS class.

2.1 Liveness and reversibility analysis of process-resource nets with acyclic, quasi-live and strongly reversible process subnets

We start with three lemmas that are necessary for establishing the main results of this section.

LEMMA 5 *A process net $\mathcal{N}_P$ satisfying Assumptions 2 and 3, is live.*

Proof: Consider a process net $\mathcal{N}_P = (P, T, W, M_0)$ and any marking $M \in R(\mathcal{N}_P, M_0)$. The net reversibility presumed by Assumption 3 implies that the net initial marking M_0 is accessible by marking M. But then, the net quasi-liveness presumed by Assumption 2 implies that any transition t is live at M. ◊

LEMMA 6 *Consider a process-resource net $\mathcal{N} = (P_S \cup I \cup O \cup P_R, T, W, M_0)$ with quasi-live and strongly reversible process subnets. If there exists a marking $M \in R(\mathcal{N}, M_0)$ s.t. there is a process subnet $\mathcal{N}_{P_{j*}}$ with $M(i_{j*}) + M(o_{j*}) \neq M_0(i_{j*})$ and $\overline{M}$ is a total deadlock, then there exists a siphon S s.t.*

i S is deadly marked at $\overline{M}$;

ii $S \cap P_R \neq \emptyset$;

iii $\forall p \in S \cap P_R$, *p is a disabling place at $\overline{M}$.*

Proof: Let S denote the set of disabling places in modified marking $\overline{M}$. Since $\overline{M}$ is a total deadlock, $S^\bullet = T \supseteq {}^\bullet S$. Therefore, S is a siphon, and the definition of S also implies that it is deadly marked. This establishes part (i) in the above lemma.

To establish that $S \cap P_R \neq \emptyset$, consider the process subnet $\mathcal{N}_{P_{j^*}}$. The fact that $M(i_{j^*}) + M(o_{j^*}) \neq M_0(i_{j^*})$ implies that there are active process instances in the subnet $\mathcal{N}_{P_{j^*}}$. But then, Assumptions 2 and 3 imply that subnet $\mathcal{N}_{P_{j^*}}$ remains live in spite of any token removal from places i_{j^*} and o_{j^*} requested by Definition 25. Hence, the occurrence of the system deadlock at $\overline{M}$ must involve insufficiently marked resource places.

Finally, part (iii) of Lemma 6 is an immediate consequence of the above definition of set S. $\diamond$

LEMMA 7 *Consider a process-resource net $\mathcal{N} = (P_S \cup I \cup O \cup P_R, T, W, M_0)$ with acyclic, quasi-live and strongly reversible process subnets. If $\mathcal{N}$ is not live, then, there exists a marking $M \in R(\mathcal{N}, M_0)$ s.t. (i) there is a process subnet $\mathcal{N}_{P_{j^*}}$ with $M(i_{j^*}) + M(o_{j^*}) \neq M_0(i_{j^*})$ and (ii) $\overline{M}$ is a total deadlock.*

Proof: Since $\mathcal{N}$ is not live, there exists a marking $M' \in R(\mathcal{N}, M_0)$ and a transition $t' \in T$ s.t. t' is dead in M'. We claim that there also exists a marking $M \in R(\mathcal{N}, M')$ s.t. (i) there exists a process subnet $\mathcal{N}_{P_{j^*}}$ with $M(i_{j^*}) + M(o_{j^*}) \neq M_0(i_{j^*})$ and (ii) every transition $t \notin (I \cup O)^\bullet$ is disabled in M. Indeed, the acyclic structure of the process subnets $\mathcal{N}_{P_j}$, $j = 1, \ldots, n$, implies that every transition sequence σ s.t. $M'[\sigma\rangle$ and for all $t \in \sigma$, $t \notin (I \cup O)^\bullet$, will be of finite length. Consider such a maximal transition sequence $\hat{\sigma}$ and let $M'[\hat{\sigma}\rangle M$. Then, at marking M there must be a process subnet $\mathcal{N}_{P_{j^*}}$ with $M(i_{j^*}) + M(o_{j^*}) \neq M_0(i_{j^*})$, since otherwise the initial marking M_0 is reachable from M, and the quasi-liveness of $\mathcal{N}$ implies that t' is not dead at M'. $\overline{M}$ is a total deadlock for $\mathcal{N}$ since the specification of $\overline{M}$, by setting $\overline{M}(i_j) = \overline{M}(o_j) = 0$, $\forall j$, essentially disables all transitions $t \in (I \cup O)^\bullet$, that, by construction, are the only transitions potentially enabled in M. $\diamond$

In the following, a deadly marked siphon S that satisfies the conditions (ii) and (iii) of Lemma 6, will be called a *resource-induced* deadly marked siphon. Lemma 6 essentially specializes the more general connection between total deadlocks and deadly marked siphons (c.f. Theorem 5.1 of Section 1.1), to the subclass of process-resource nets with quasi-live and strongly reversible active processes. From a methodological standpoint, it provides a vehicle for connecting the liveness – and, in certain cases, even the quasi-liveness – of the resource allocation taking place in process-resource nets, to resource-induced deadly marked siphons, as long as it can be established that the lack of (any of) these properties implies the existence of a reachable marking M s.t. (i) there exists a process subnet $\mathcal{N}_{P_{j*}}$ with $M(i_{j*}) + M(o_{j*}) \neq M_0(i_{j*})$ and (ii) the corresponding modified marking $\overline{M}$ is a total deadlock. Lemma 7 establishes that this is the case for the class of process-resource nets with acyclic, quasi-live and strongly reversible process subnets. The next theorem completes the characterization of the relationship between the non-liveness in the class of process-resource nets with acyclic, quasi-live and strongly reversible process subnets, and the presence of resource-induced deadly marked siphons in the net modified reachability space.

THEOREM 5.3 *Let* $\mathcal{N} = (P_S \cup I \cup O \cup P_R, T, W, M_0)$ *be a process-resource net with acyclic, quasi-live and strongly reversible processes.* $\mathcal{N}$ *is live if and only if the space of modified reachable markings,* $\overline{R(\mathcal{N}, M_0)}$, *contains no resource-induced* deadly marked siphons.

Proof: To show the necessity part, suppose that there exists a marking $M \in R(\mathcal{N}, M_0)$ s.t. $\overline{M}$ contains a resource-induced deadly marked siphon S. Let $r \in S \cap P_R$ be one of the disabling resource places, and consider $t \in r^{\bullet}$ s.t. $\overline{M}(r) < W(r, t)$. The definition of deadly marked siphon implies that all transitions $t' \in {}^{\bullet}r$ are dead in $R(\mathcal{N}, \overline{M})$. This remark, when combined with Definition 25 and Assumption 4, further imply that for all markings $M' \in R(\mathcal{N}, M)$, $M'(r) \leq M(r)$, since the re-introduction of the tokens removed from places $p \in I \cup O$ and their potential loading in the system, can only decrease the resource availabilities. Therefore, t is a dead transition at M, which contradicts the assumption of net liveness.

To show the sufficiency part, suppose that $\mathcal{N}$ is not live. Then, Lemma 7 implies that there exists a marking $M \in R(\mathcal{N}, M_0)$ s.t. (i) there is a process subnet $\mathcal{N}_{P_{j*}}$ with $M(i_{j*}) + M(o_{j*}) \neq M_0(i_{j*})$, and (ii) $\overline{M}$ is a total deadlock. But then, Lemma 6 implies that $\overline{R(\mathcal{N}, M_0)}$ contains a resource-induced deadly marked siphon, which contradicts the working hypothesis. ◇

The next corollary of Theorem 5.3 is a rather technical result as it establishes that for the case of process-resource nets belonging to the class of *PT-ordinary* PN's, the problematic siphons interpreting the RAS non-liveness are, in fact, *empty* siphons, and they can also be identified in the *original* net reachability space $R(\mathcal{N}, M_0)$ (besides the modified reachability space $\overline{R(\mathcal{N}, M_0)}$).

COROLLARY 3 *Let* $\mathcal{N} = (P_S \cup I \cup O \cup P_R, T, W, M_0)$ *be a* PT-ordinary *process-resource net with acyclic, quasi-live and strongly reversible process subnets.* $\mathcal{N}$ *is live if and only if the space of reachable markings,* $R(\mathcal{N}, M_0)$, *contains no empty siphons.*

Proof: According to Theorem 5.3, under the assumptions of Corollary 3, net $\mathcal{N}$ is not live *iff* there exists a marking $M \in R(\mathcal{N}, M_0)$, s.t. $M \neq M_0$ and its modified marking $\overline{M}$ contains a resource-induced deadly marked siphon, S. Furthermore, the development of the result of Theorem 5.3 (c.f., the proofs of Lemmas 6 and 7) indicates that S is defined by the set of disabling places of a total deadlock contained in $\overline{M}$. Since every place $p \in S$ is a disabling place in $\overline{M}$, and net $\mathcal{N}$ is PT-ordinary, $\overline{M}(p) = 0$, $\forall p \in S$. Hence, S is an empty siphon in $\overline{M}$. It remains to be shown that the presence of the resource-induced empty siphon S in the modified marking $\overline{M}$ implies the presence of an empty siphon S' in the original marking M. For that, let $S' = \{r_i : r_i \in S\} \cup \{p \in P_S : M(p) = \overline{M}(p) = 0 \wedge \exists r_i \text{ s.t. } (r_i \in S \wedge y_{r_i}(p) > 0)\}$. Notice that $S' \neq \emptyset$, since S is a resource-induced empty siphon. We show that S' is a siphon (which is empty, by construction), by considering the next two main cases:

Case I – $t \in \,^{\bullet}r_k$ for some $r_k \in S$: Then, $\exists q \in S$ s.t. $t \in q^{\bullet}$. If $q \in P_R$, then $q \in \{r_i : r_i \in S\} \subset S'$. On the other hand, if $q \notin P_R$, then $q \in P_S$, since $(q^{\bullet})^{\bullet} \cap P_R \neq \emptyset$. Furthermore, $y_{r_k}(q) > 0$ and $M(q) = 0$ (since $q \in S$). Therefore, $q \in \{p \in P_S : M(p) = \overline{M}(p) = 0 \wedge \exists r_i \text{ s.t. } (r_i \in S \wedge y_{r_i}(p) > 0)\} \subset S'$. In both cases, $t \in (S')^{\bullet}$.

Case II – $t \in \,^{\bullet}q$ for some $q \in P_S$ with $M(q) = \overline{M}(q) = 0 \wedge \exists r_k \text{ s.t. } (r_k \in S \wedge y_{r_k}(q) > 0)$: Then, if $\exists r_l$ s.t. $r_l \in S \wedge t \in r_l^{\bullet}$, $t \in \{r_i : r_i \in S\}^{\bullet} \subseteq (S')^{\bullet}$. Otherwise, $\exists q' \in (I \cup O \cup P_S) \cap \,^{\bullet}t$ with $\overline{M}(q') = 0$. Furthermore, since $y_{r_k}(q) > 0$ and, by the sub-case assumption, $\forall r_l \in \,^{\bullet}t$, $M(r_l) > 0$, it must be that $y_{r_k}(q') > 0$. But then, $t \in \{p \in P_S : M(p) = \overline{M}(p) = 0 \wedge \exists r_i \text{ s.t. } (r_i \in S \wedge y_{r_i}(p) > 0)\}^{\bullet} \subseteq (S')^{\bullet}$. $\diamond$

The last result of this section establishes that, in the considered class of process-resource nets, liveness and reversibility are equivalent concepts.

THEOREM 5.4 *A process-resource net* $\mathcal{N} = (P_S \cup I \cup O \cup P_R, T, W, M_0)$ *with acyclic, quasi-live, and strongly reversible process subnets is reversible if and only if it is live.*

Proof: The necessity ("only-if") part of this theorem results immediately from Lemma 7, since otherwise there exists a marking $M \in R(\mathcal{N}, M_0)$ s.t. there is a process subnet $\mathcal{N}_{P_{j*}}$ with $M(i_{j*}) + M(o_{j*}) \neq M_0(i_{j*})$, and $\overline{M}$ is a total deadlock.

In order to establish the sufficiency ("if") part of the theorem, consider a marking $M \in R(\mathcal{N}, M_0)$ s.t. $M \neq M_0$. Then, we discern two cases: In the first case, every process subnet $\mathcal{N}_{P_j}$ has $M(i_j) + M(o_j) = M_0(i_j)$, and it should be obvious from the structure of net $\mathcal{N}$, that $M_0 \in R(\mathcal{N}, M)$. In the alternative case, using an argument similar to that in the proof of Lemma 7, one can construct a maximal-length firing sequence σ leading to a marking M' s.t. every transition $t \notin (I \cup O)^{\bullet}$ is disabled in M'. We claim that at M', every process subnet $\mathcal{N}_{P_j}$ has $M'(i_j) + M'(o_j) = M_0(i_j)$, and therefore, $M_0 \in R(\mathcal{N}, M')$, which further implies that $M_0 \in R(\mathcal{N}, M)$. Indeed, by

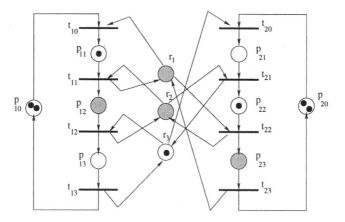

Figure 5.7. Example I: Interpreting the non-liveness of PT-ordinary process-resource nets with acyclic, quasi-live and strongly reversible process subnets through empty siphons

construction, $\overline{M'}$ is a total deadlock of $\mathcal{N}$, and if there is a process subnet $\mathcal{N}_{P_{j*}}$ with $M'(i_{j*}) + M'(o_{j*}) \neq M_0(i_{j*})$, Lemma 6 implies that M' contains a resource-induced deadly marked siphon. But then, Theorem 5.3 implies that $\mathcal{N}$ is not live, which contradicts the working hypothesis. ◇

Examples

The next series of examples demonstrates the results derived in this section. It also seeks to provide a more intuitive interpretation of the previous theoretical developments.

I. Siphon-based interpretation of the non-liveness of PT-ordinary process-resource nets with acyclic, quasi-live and strongly reversible process subnets. A careful study of the proof of Corollary 3 will reveal that the empty siphons S' underlying the non-liveness of PT-ordinary process-resource nets with acyclic, quasi-live and strongly reversible process subnets can be identified in markings M that correspond to RAS states in which only deadlocked processes are loaded in the system. In addition, these siphons consist of (i) resource places $r_l \in P_R$ corresponding to the resources entangled into deadlock and (ii) places $p \in P_S$ corresponding to processing stages that are supported by the aforementioned resources but are unmarked in marking M; this second set of places are known as *empty holders* in the relevant literature.

The structure of siphon S' is demonstrated in Figure 5.7. The process-resource net depicted in Figure 5.7 is the same with the net depicted in Figure 5.6, but the depicted marking corresponds to a deadlock state of the underlying RAS: a process instance executing stage Ξ_{11} holds the only unit of resource R_1 and requests allocation of the single unit of resource R_2, which, however, is held by a process instance executing stage Ξ_{22} and requesting the allocation of resource R_1. Hence, in this case, the resources entangled into deadlock are R_1 and R_2, and their empty holders are the places p_{12} and p_{23}, corresponding to stages Ξ_{12} and Ξ_{13}. Therefore, the empty siphon S' identified in the proof of Corollary 3, consists of places r_1, r_2, p_{12} and p_{23}, that are shaded in Figure 5.7. Indeed, the reader can verify that $^\bullet S \subseteq S^\bullet$; in particular, $^\bullet S = \{t_{11}, t_{23}, t_{12}, t_{22}\}$ and $S^\bullet = \{t_{10}, t_{22}, t_{11}, t_{21}, t_{12}, t_{23}\}$. $\diamond$

II. Siphon-based interpretation of the non-liveness of non-PT-ordinary process-resource nets with acyclic, quasi-live and strongly reversible process subnets. The concept of the empty siphon S' discussed in the previous example will fail to explain the non-liveness of non-PT-ordinary process-resource nets with acyclic, quasi-live and strongly reversible process subnets. This is a consequence of the *non-uniformity* of the resource allocation requests posed by the various processing stages with respect to any single resource type, which allows the existence of RAS states where certain process types can be repeatedly executed, even though some of their supporting resource types are involved in a deadlock. The situation is depicted in Figures 5.8-5.10: Figure 5.9 depicts a reachable marking of the process-resource net with acyclic, quasi-live and strongly reversible process subnets depicted in Figure 5.8, where the active process instances corresponding to the tokens in place p_{21} are deadlocked, since their request for 2 extra units of resource R_1 cannot be met unless one of them releases its currently held resource. However, process instances executing stage p_{11} can still engage the remaining free unit of resource R_1 and successfully proceed to completion. But then, place p_{11} cannot be part of an empty siphon, even though it is an empty holder of resource R_1, that is involved in the depicted RAS deadlock.

The aforementioned problem is remedied by the introduction of the concept of the *modified marking*. The modified marking of the original marking depicted in Figure 5.9 is depicted in Figure 5.10. By removing all the tokens resident in places i_j, $\forall j$, we construct a deadlock marking, in which, according to Theorem 5.1, the set S of all disabling places – depicted by shaded places in Figure 5.10 – will be a deadly marked siphon. It is interesting to notice that the constructed siphon S is deadly marked but not empty. Furthermore, notice that the token removal implied by the definition of the modified marking $\overline{M}$ will, in general, generate artificially empty siphons, especially, for those process instances with no active processes in the original marking M; in Figure 5.10, $S' = \{p_{10}, p_{11}\}$ is such an artificially constructed empty siphon. It is due to this effect that the result of Theorem 5.3 stipulates the presence of *resource-induced* deadly marked siphons – i.e., deadly marked siphons containing disabling resource places – in order to infer the net non-liveness. $\diamond$

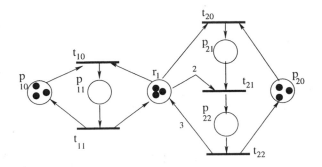

Figure 5.8.   Example II: The considered process-resource net

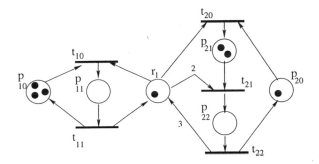

Figure 5.9.   Example II: A net marking containing a RAS deadlock

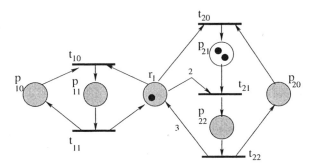

Figure 5.10.   Example II: The resource-induced deadly marked siphon

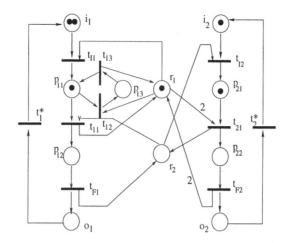

Figure 5.11. A case of RAS non-liveness that cannot be attributed to the development of resource-induced deadly marked siphons, due to the presence of enabled internal process cycles

III. The importance of process net acyclicity for the result of Theorem 5.3.

It should be clear to the reader that Assumption 1 is instrumental for establishing the validity of Theorem 5.3. Indeed, the example net of Figure 5.11 demonstrates that for the case of process-resource nets where the process flows can present internal cycles, the structural concept of resource-induced deadly marked siphon is not sufficient for interpreting the net non-liveness, even under the assumptions of quasi-live and strongly reversible process subnets.[3] The RAS depicted in Figure 5.11 has two resource types, R_1 and R_2, respectively available at 2 and 1 units, and two process types, Π_1 and Π_2. Process type Π_1 involves three stages, Ξ_{11}, Ξ_{12} and Ξ_{13}, with corresponding resource allocation requests $(1,0)^T$, $(0,1)^T$ and $(2,0)^T$. Process type Π_2 involves two stages, Ξ_{21} and Ξ_{22}, with corresponding resource allocation requests $(0,1)^T$ and $(2,0)^T$. The complete flow logic for these two process types is specified by the process subnets depicted in Figure 5.11. The marking $M \in R(\mathcal{N}, M_0)$ depicted in Figure 5.11 corresponds to a RAS deadlock involving the two active processes (transitions t_{11} and t_{21} are dead in the depicted marking M). Yet, the corresponding modified marking $\overline{M}$, as well as all the modified markings $\overline{M'} \in \overline{R(\mathcal{N}, \overline{M})}$, contain no resource-induced deadly marked siphon. This ab-

---

[3]The works of (Park, 2000; Jeng and Xie, 2001) identify some special structure on the resource allocation requests that allows the attribution of the net non-liveness to resource-induced deadly marked siphons, even for the case of process-resource nets with cyclic process routes. However, the complete characterization of the dynamics and the liveness-related properties of process-resource nets with cyclic process routes is an issue open to further investigation.

Input: A marked PN $\mathcal{N} = (P, T, W, M_0)$ and a marking $M \in R(\mathcal{N}, M_0)$
Output: The maximal deadly marked siphon in M, S

1　$S := P$;　$\mathcal{N}' := \mathcal{N}$

2　**while** $\exists\, t \in T$ such that t is fireable in the modified net $\mathcal{N}'$ **do**

　　(a)　$S := S \backslash t^\bullet$

　　(b)　Remove t from $\mathcal{N}'$

　　(c)　Remove $t^\bullet$ from $\mathcal{N}'$

　　endwhile

3　**Return** S

Figure 5.12.　An algorithm for computing the maximal deadly marked siphon in a given PN marking M

sence of a deadly marked siphon can be explained by the fact that the deadlocked process in place p_{11} can circulate freely in the circuit $< p_{11}, t_{12}, p_{13}, t_{13} >$, and therefore, the $\overline{\text{RAS}}$ deadlock of the two processes does not translate to a total deadlock in $\overline{R(\mathcal{N}, \overline{M})}$. ◇

2.2　Liveness verification for process-resource nets with acyclic, quasi-live and strongly reversible process subnets

This section exploits the siphon-based characterization of non-liveness for process-resource nets with acyclic, quasi-live, and strongly reversible process subnets, in order to develop an analytical test for verifying the liveness of any given instance of the considered PN sub-class. The derived test takes the form of a *mixed integer programming (MIP)* formulation, that is polynomially sized, in terms of variables and constraints, with respect to the underlying PN model. As a result, it is very practical from a computational standpoint. The practical value of this test is further demonstrated in the next section where it is shown how it can be integrated in a synthesis procedure that supports the design of *algebraic* SCP's for the considered class of process-resource nets.

The starting point for the development of the aforementioned MIP formulation is the observation that, given a marked PN $\mathcal{N} = (P, T, W, M_0)$ and a marking $M \in R(\mathcal{N}, M_0)$, the maximal deadly marked siphon S in M can be computed by the algorithm of Figure 5.12, originally developed in (Park and Reveliotis, 2001b). The next theorem formally states and proves this result.

THEOREM 5.5 *Given a PN $\mathcal{N} = (P, T, W, M_0)$ and a marking $M \in R(\mathcal{N}, M_0)$, the algorithm of Figure 5.12 returns the maximal deadly marked siphon S in M.*

Proof: Consider the place set S returned by the algorithm of Figure 5.12, and let t denote some transition in ${}^\bullet S$. t does not belong to the set of transitions removed from the net $\mathcal{N}$ during the algorithm execution, since, otherwise, $t^\bullet \not\subseteq S$. But then, t is a transition not fireable in $\mathcal{N}$, due to the insufficient marking of some place(s) $p \in S$. Hence, $t \in S^\bullet$, and S is a siphon that is deadly marked in M.

Next, we show that S is the maximal deadly marked siphon in M, by using contradiction. Suppose that S' is another deadly marked siphon in M with $S' \backslash S \neq \emptyset$, and let $p \in S' \backslash S$. Since p is removed from net $\mathcal{N}$ by the considered algorithm, there exists transition $t \in {}^\bullet p \subseteq {}^\bullet S'$ that either (i) is fireable in M, or (ii) it has all of its input places removed from $\mathcal{N}$ during the algorithm execution. Case (i) is not possible since S' would not be a deadly marked siphon in M. Therefore, Case (ii) must apply, and there must exist another place $p' \in S' \backslash S$ that has been removed from net $\mathcal{N}$ by the algorithm prior to the removal of place p (since S' is a siphon and $t \in {}^\bullet S'$). But then, the entire above argument can be applied to place p' leading to the identification of another place $p'' \in S' \backslash S$, that has been removed from the algorithm before the places p and p', which, in turn, can lead to the identification of another similar place p''', etc. The existence of such a place sequence is contradicted by the finiteness of the place set P. $\diamond$

For the case of *structurally bounded* nets, the algorithm of Figure 5.12 can be converted to a MIP formulation as follows: First, let $SB(p)$ denote a structural bound for the markings of place $p \in P$. Furthermore, let v_p, z_t and f_{tp} be *binary indicator* variables respectively denoting the following conditions:

$$v_p = 1 \iff \text{place } p \text{ is removed by the algorithm, } \forall p \in P$$
$$z_t = 1 \iff \text{transition } t \text{ is removed by the algorithm, } \forall t \in T$$
$$f_{pt} = 1 \iff M(p) \geq W(p,t) \vee v_p = 1, \ \forall W(p,t) > 0$$

Then, we have the following theorem:

THEOREM 5.6 *Given a marking $M \in R(\mathcal{N}, M_0)$ of a structurally bounded PN $\mathcal{N} = (P, T, W, M_0)$, the maximal deadly marked siphon S contained in M is determined by:*

$$S = \{p \in P \mid v_p = 0\} \tag{5.5}$$

where v_p, $p \in P$, is obtained through the following IP formulation:

$$G(M) = \min \sum_{p \in P} v_p \tag{5.6}$$

s.t.

$$f_{pt} \geq \frac{M(p) - W(p,t) + 1}{SB(p)}, \quad \forall W(p,t) > 0 \tag{5.7}$$

$$f_{pt} \geq v_p, \quad \forall W(p,t) > 0 \tag{5.8}$$

$$z_t \geq \sum_{p \in {}^{\bullet}t} f_{pt} - |{}^{\bullet}t| + 1, \quad \forall t \in T \tag{5.9}$$

$$v_p \geq z_t, \quad \forall W(t,p) > 0 \tag{5.10}$$

$$v_p, z_t, f_{pt} \in \{0,1\}, \quad \forall p \in P, \forall t \in T \tag{5.11}$$

Proof: Equation 5.9 together with Equation 5.7 imply that all transitions z_t fireable in marking M will have $z_t = 1$. Furthermore, Equation 5.10 implies that all places $p \in t^{\bullet}$ for some t with $z_t = 1$ will have $v_p = 1$, which implements Step (2.b) in the algorithm of Figure 5.12. Similarly, Equation 5.8 combined with Equation 5.9 force $z_t = 1$ for all transitions t with $v_p = 1$, $\forall p \in {}^{\bullet}t$. Finally, the fact that no additional place p (resp., transition t) has $v_p = 1$ (resp., $z_t = 1$), is guaranteed by the specification of the objective function in the above formulation. $\diamond$

In the case that net $\mathcal{N}$ is a process-resource net, the formulation of Theorem 5.6 can be restricted to the computation of the maximal *resource-induced* deadly marked siphon, through the introduction of the following two constraints (Park and Reveliotis, 2001b):

$$\sum_{r \in P_R} v_r \leq |P_R| - 1 \tag{5.12}$$

$$\sum_{t \in r^{\bullet}} f_{rt} - |r^{\bullet}| + 1 \leq v_r, \quad \forall r \in P_R \tag{5.13}$$

Constraint 5.12 enforces that the identified siphon S must contain at least one resource place, while Constraint 5.13 requires that all resource places included in S must be disabling. The resulting necessary and sufficient condition for the non-existence of resource-induced deadly marked siphons in a given marking M of a process-resource net is as follows:

COROLLARY 4 *A given marking M of a process-resource net $\mathcal{N}$ contains no resource-induced deadly marked siphons, if and only if the corresponding formulation of Equations 5.6–5.13 is infeasible.*

Proof: Theorem 5.6 implies that the set of places $S' = \{p \in P : v_p = 0\}$, satisfying the formulation of Equations 5.6-5.11 is the maximal deadly marked siphon at M. Next we show that Equation 5.13 eliminates from the set S' all places $r \in S' \cap P_R$ that do not disable any transition $t \in r^\bullet$, while the remaining set of places, S, maintains the deadly marked siphon property. Indeed, a place $r \in S' \cap P_R$ for which $v_r = 0$ in the solution computed by the IP formulation of Theorem 5.6, will have $v_r = 1$ under the addition of Constraint 5.13, only if $\sum_{t \in r^\bullet} f_{rt} = |r^\bullet|$ (i.e., only if $M(r)$ disables no output transition of r). Furthermore, the "removal" of such a place r, by setting $v_r = 1$, does not incur the "removal" of any additional places and transitions, because all the transitions t with $z_t = 0$ in the solution of the IP formulation of Theorem 5.6, are disabled in marking M; i.e., for every $t \in r^\bullet$ there exists an additional place $p \in {}^\bullet t \backslash \{r\}$ s.t. $M(p, t) < W(p, t)$. The last observation implies that the set of places $S \equiv S' \backslash \{r : r \in S' \cap P_R \wedge v_r = 1$ by Equation 5.13$\}$ is a deadly marked siphon (since for all transitions $t \in {}^\bullet S$, there exists a place $p \in S$ s.t. $t \in p^\bullet$ and p disables t at M). In addition, since S' is maximal, S is also maximal. Finally, since Equation 5.12 requires that $S \cap P_R \neq \emptyset$, it will be infeasible if and only if the considered marking M contains no resource-induced deadly marked siphons. $\diamond$

The test of Corollary 4 can be extended, in principle, to a test for the non-existence of resource-induced deadly marked siphons over the entire modified reachability space, $\overline{R(\mathcal{N}, M_0)}$, of a process-resource net $\mathcal{N} = (P, T, W, M_0)$, by:

i substituting marking vector M in the MIP formulation of Theorem 5.6 with the modified marking vector $\overline{M}$;

ii introducing an additional set of variables, M, representing the net reachable markings;

iii adding two sets of constraints, the first one linking variables M and $\overline{M}$ according to the logic of Equation 5.4, and the second one ensuring that the set of feasible values for the variable vector M is equivalent to the PN reachability space $R(\mathcal{N}, M_0)$.

Unfortunately, any system of linear inequalities characterizing exactly the set $R(\mathcal{N}, M_0)$ is of exponential complexity with respect to the net size (Silva et al., 1998). On the other hand, a superset of the reachability space $R(\mathcal{N}, M_0)$ is provided by the system *state equations* 5.2-5.3. These remarks lead to a *sufficient* condition for the non-existence of resource-induced deadly marked siphons S in the entire space $\overline{R(\mathcal{N}, M_0)}$ of a given process-resource net $\mathcal{N}$, which in the light of Theorem 5.3, constitutes a *sufficient* condition for the live-ness of process-resource nets with acyclic, quasi-live, and strongly reversible process subnets.

COROLLARY 5 *Let $\mathcal{N} = (P, T, W, M_0)$ be a process-resource net with acyclic, quasi-live, and strongly reversible process subnets. If the mixed integer pro-*

gram defined by (i) Equations 5.6–5.13, where vector variable M is replaced by vector variable $\overline{M}$, (ii) Equations 5.2–5.3, and (iii) Equation 5.4, is infeasible, then $\mathcal{N}$ is live.

Concluding this section, we notice that for the case of PT-ordinary process-resource nets with acyclic, quasi-live, and strongly reversible subnets, a similar but simpler, from a computational standpoint, liveness sufficiency test can be derived, based on the result of Corollary 3. We refer the reader to (Chu and Xie, 1997; Park and Reveliotis, 2000) for a detailed discussion of the corresponding formulation.

3.   Liveness-enforcing supervision of process-resource nets with acyclic, quasi-live and strongly reversible process subnets

In this section we discuss two approaches for developing correct SCP's – or, in the spirit of Theorem 5.4, *Liveness-Enforcing Supervisors (LES)* – for the class of RAS with complex process flows considered in this chapter. The first approach capitalizes on the results of Section 2.2, as it seeks to augment the PN modelling the underlying resource allocation with a control subnet, such that the resulting controlled net is a live and reversible process-resource net with acyclic, quasi-live and strongly reversible process subnets. The second approach seeks to extend and exploit the results of Chapter 4, by (i) using these results in order to develop PK-SCP's for the class of DIS-CON-RAS, and subsequently (ii) reducing any more complex RAS behavior to a DIS-CON-RAS model that can be controlled by the developed SCP's. In order to remain consistent with the design specifications posed in Section 3 of Chapter 1, the developed policies must have a representation and a "run-time" computational cost that are polynomially related to the underlying RAS size, or, equivalently, to the size of the RAS-modelling PN. On the other hand, we shall allow the task of developing such a policy for any given RAS configuration to be of non-polynomial complexity with respect to the considered RAS size, since, typically, this task will be performed *"off-line"*.

3.1   Synthesizing correct algebraic PK-SCP's for process-resource nets with acyclic, quasi-live and strongly reversible process subnets

The concept of *algebraic* SCP's was introduced in Chapter 4, as the class of SCP's that can be expressed by a set of linear inequalities to be satisfied by the RAS state. Furthermore, in order to remain in the domain of PK-SCP's, this set of inequalities must be polynomially sized with respect to the size of the underlying RAS. In the RAS representation of process-resource nets, the number of active process instances at any given processing stage is expressed by the marking of the corresponding place $p \in P_S$. Therefore, an algebraic PK-SCP $A \cdot s \leq b$ is expressed in the PN modelling framework by the equation

$$A \cdot M_S \leq b \qquad (5.14)$$

where M_S restricts the marking M of the RAS-modelling PN to its components corresponding to places $p \in P_S$. The constraints expressed by Equation 5.14 can subsequently be enforced in the PN-based representation of the RAS dynamics through the super-imposition on the original process-resource net of a controlling subnet that is readily constructed through the theory of *control-place invariants*, presented in (Moody and Antsaklis, 1998). According to (Moody and Antsaklis, 1998), each of the inequalities

$$A_{[l,\cdot]} \cdot M_S \leq b_l \qquad (5.15)$$

can be imposed on the net behavior by superimposing on the original net structure a *control* place w_l, connected to the rest of the network according to the flow matrix

$$\theta_{w_l} = -A_{[l,\cdot]} \cdot \Theta_S \qquad (5.16)$$

where Θ_S denotes the flow sub-matrix of the uncontrolled network $\mathcal{N} = (P, T, W, M_0)$ corresponding to places $p \in P_S$. The initial marking of place w_l must be set to

$$M_0(w_l) = b_l \qquad (5.17)$$

and the resulting controller imposes Constraint 5.15 on the system behavior by establishing the place invariant

$$A_{[l,\cdot]} \cdot M_S + M(w_l) = b_l \qquad (5.18)$$

Equation 5.18, when interpreted in the light of Assumption 4 of Section 1.2, implies that the control places w_l, implementing each of the constraints in the LES-defining Equation 5.14, essentially play the role of fictitious new resources in the dynamics of the net $\mathcal{N}^c$, that models the controlled system behavior. As a result, the controlled net $\mathcal{N}^c$ remains in the class of process-resource nets that satisfy Assumptions 1 and 3. Let $P_W \equiv \bigcup_l \{w_l\}$. If it can be shown that the net $\mathcal{N}^c$ satisfies also Assumption 5 with respect to the extended "resource" set $P_R \cup P_W$, it can be inferred that $\mathcal{N}^c$ belongs to the class of process-resource nets with acyclic, quasi-live and strongly reversible process subnets, and therefore, the correctness of the considered algebraic SCP can be verified through the liveness sufficiency test of Corollary 5, applied on the controlled net $\mathcal{N}^c$.

In the general case, the assessment of the quasi-liveness of the various resource-augmented process subnets of the controlled net $\mathcal{N}^c$ can be performed through standard reachability analysis techniques available in the PN literature (Cassandras and Lafortune, 1999). These techniques are of super-polynomial complexity with respect to the size of the involved process subnets, but they will be performed off-line. In addition, they will be executed separately on each process net, a decomposition that decreases drastically the size of the deployed state spaces. Furthermore, in certain cases, special structure can be exploited for the development of more expedient tests. For instance, as it was indicated in Section 1.2, assessing the quasi-liveness of DIS-RAS is a very straightforward task. On the other hand, even though quasi-liveness is an NP-hard problem for the class of COR-RAS (c.f., Theorem 5.2), it can be connected to the development of resource-induced deadly marked siphons, by a result similar to that of Theorem 5.3, and therefore, it can be tested through application of the test of Corollary 5 on each resource-augmented process subnet; we refer to (Reveliotis, 2003b) for a complete development of the relevant results. Another case that admits easy resolution of the quasi-liveness of the controlled net $\mathcal{N}^c$, is the case where the applied supervisor seeks only to restrict the number of process instances that are loaded simultaneously into the system, rather than their access to particular segments of their process routes. This kind of control logic is more generally known as *"process-release"* control in the resource allocation literature. Algebraic SCP's implementing process-release control policies can be expressed by a single linear inequality

$$a \cdot M_S \leq b \qquad\qquad (5.19)$$

where b defines the ceiling on the process concurrency imposed by the considered supervisor, and the elements of the row vector a are provided by a set of p-semiflows characterizing the execution flow logic of the various process types. The LES of Equation 5.19 is superimposed to the original process-resource net $\mathcal{N}$ through the introduction of a single control place w with $w^\bullet \subseteq \bigcup_j \{t_{I_j}\}$. Hence, it should be obvious to the reader that the resulting controlled net $\mathcal{N}^c$ preserves the quasi-liveness of the original net $\mathcal{N}$, as long as $M_0(w) \geq 1$. The following example demonstrates the SCP synthesis method discussed in this section, and also, the particular concept of process-release control.

Example. Consider the process-resource net depicted in Figure 5.13. As it can be seen from the figure, the underlying RAS consists of two process types, Π_1 and Π_2, and five resource types, $R_1, \ldots, R_5$. Process type Π_1 has a flow represented by an acyclic marked graph, and it involves six processing stages, $\Xi_{11}, \ldots, \Xi_{16}$, with corresponding resource requirements: $(1, 0, 0, 0, 0)^T$, $(0, 1, 0, 0, 0)^T$, $(0, 0, 1, 0, 0)^T$, $(0, 0, 1, 0, 0)^T$, $(0, 1, 0, 0, 0)^T$ and $(0, 0, 0, 0, 1)^T$. Process type Π_2 has a flow represented by an acyclic state machine, and it involves four stages, $\Xi_{21}, \ldots, \Xi_{24}$, with corresponding resource requirements: $(0, 1, 0, 0, 0)^T$, $(1, 1, 0, 0, 0)^T$, $(0, 1, 1, 0, 0)^T$ and $(0, 0, 0, 1, 0)^T$. A closer inspection of the stage resource requirements for these two processes reveals that the only resources that could be entangled in a deadlock are R_1, R_2 and R_3.

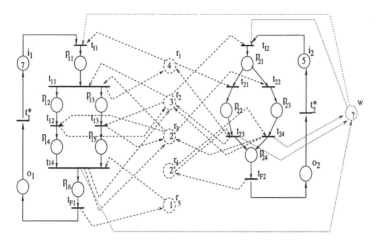

Figure 5.13. Example: The considered process-resource net and the imposed process-release control policy

Therefore, the critical sections for Π_1 and Π_2 are respectively defined by the stage sets $\{\Xi_{11}, \Xi_{12}, \Xi_{13}, \Xi_{14}, \Xi_{15}\}$ and $\{\Xi_{21}, \Xi_{22}, \Xi_{23}\}$.

Our intention is to develop an LES that will establish the liveness of the controlled net by restricting the number of process instances that can simultaneously execute in their critical sections identified above. Hence, the proposed supervisor constitutes a more refined implementation of the general *"process-release"* control scheme, to the particular process-resource net of Figure 5.13. This LES is super-imposed to the original process-resource net of Figure 5.13 by introducing the control place w, connected to the original process-resource net through the flow structure depicted by dotted lines in Figure 5.13.

Next we seek to determine the maximal initial marking for place w that leads to live behavior for the controlled net structure of Figure 5.13, using the siphon-based liveness analysis suggested by the results of Section 2.2. For this, we first determine an upper bound to the maximal number of processes that can be executed simultaneously by the considered RAS. The reader can convince herself that, based on the resource capacities and the process flows annotated in Figure 5.13, an upper bound for the system concurrency w.r.t. process type Π_1 (resp., Π_2) is 7 (resp., 5) process instances. Then, application of the MIP formulation of Corollary 5 in a binary search over the integer set $\{1, \ldots, 12\}$, reveals that the maximal initial marking for control place w leading to a correct algebraic LES – or equivalently, the maximal number of processes that can be simultaneously loaded in the system without the possibility of running into any deadlocking problems – is 6. For completeness, we mention that the deadlock marking identified by the computerized solver when the MIP formulation of

Corollary 5 was solved with $M_0(w) = 7$, is: $M(i_1) = 1$; $M(p_{11}) = 4$; $M(p_{12}) = M(p_{13}) = 2$; $M(i_2) = 4$; $M(p_{21}) = 1$; $M(r_4) = 2$; $M(r_5) = 1$; and zero for every other place. $\diamond$

As it is demonstrated by the previous example, the detailed synthesis of the control subnet to be super-imposed on a process-resource net with acyclic, quasi-live and strongly reversible process subnets, is based on case-specific intuitive insights and arguments. In general, there is no guarantee that the resulting controlled net $\mathcal{N}^c$ will be *"structurally live"* with respect to markings $M_0(w_l)$, $l = 1, \ldots, \dim(b)$, i.e., there might be no marking $M_0(w_l)$, $l = 1, \ldots, \dim(b)$, such that the resulting net $\mathcal{N}^c$ will satisfy the liveness condition of Corollary 5. Furthermore, even though the satisfaction of the liveness condition of Corollary 5 is monotonic with respect to the marking of the control place w for the case of process-release control policies, such a monotonicity does not apply to the general class of algebraic SCP's. Therefore, there are no generally applying efficient schemes for searching the underlying marking space $M_0(w_l)$, $l = 1, \ldots, \dim(b)$. Yet, in the face of the complexity of the considered problem, and the apparent scarcity of the relevant controller design results, the presented method is a powerful policy correctness verification tool, that is further characterized by considerable design flexibility. The next section discusses another policy design approach that is more systematic in its development, but limits the scope of the considered solutions.

3.2   Synthesizing correct SCP's for CPX-CON-RAS through reduction to DIS-CON-RAS

The reduction to DIS-CON-RAS.   The LES synthesis approach discussed in this section is motivated by the following series of observations: First notice that the reachability graph of any resource-augmented process net $\overline{\mathcal{N}_P}$ with $M_0(i) = 1$ essentially constitutes a finite state automaton representing all the possible ways in which a process instance of the considered process type can execute in isolation. In the particular case that the considered resource-augmented process net is acyclic in the spirit of Assumption 1 – which is the case of interest in this book – any circuit in the STD of the aforementioned FSA will contain the markings with $M(p) = 0$, $\forall p \in P_S$, i.e., the state in which the system is idle and empty of any processes. In fact, in order to establish the deadlock-free operation of the underlying process, the behavior of any acyclic resource-augmented process net must be constrained exactly to those markings that belong to the aforementioned circuits, i.e., to the process reachable and safe subspace. Assuming that the reachable and safe subspace of the considered resource-augmented process net $\overline{\mathcal{N}_P}$ is non-trivial, – i.e., it does not contain only the markings with $M(p) = 0$, $\forall p \in P_S$ – one can replace $\overline{\mathcal{N}_P}$, for the purposes of the LES synthesis pursued in this section, with another resource-augmented process net, $\overline{\mathcal{N}'_P}$, that is defined by the strongly connected state machine corresponding to the reachable and safe subspace of $\overline{\mathcal{N}_P}$. The idle place p'_0 of this new process net aggregates the markings M of the original net with $M(p) = 0$, $\forall p \in P_S$, while its place set P'_S is defined by the set of reachable and safe markings of $\overline{\mathcal{N}_P}$ that contain active processes. The resource allocation request vector associated with each place $p' \in P'_S$ can be readily computed as the *total* resource allocation implied by the corresponding marking M of $\overline{\mathcal{N}_P}$. Furthermore, each transition $t' \in T'$ of $\overline{\mathcal{N}'_P}$ corresponds to the firing of a particular transition $t \in T$ of $\overline{\mathcal{N}_P}$, and it will be labelled by it.

From a computational standpoint, such an alternative PN-based representation of the reachable and safe subspace of any given resource-augmented process net $\overline{\mathcal{N}_P}$ can be developed by first enumerating the entire reachability space $R(\mathcal{N}_P, M_0)$ of the net $\overline{\mathcal{N}_P}$, under the assumption that $M_0(i) = 1$, and subsequently trimming $R(\mathcal{N}_P, M_0)$ with respect to the initial marking M_0. The trimming operation can be done easily by traversing backwards the reachability space $R(\mathcal{N}_P, M_0)$, starting from the initial marking M_0, and labelling all markings that are encountered in the process; unlabelled markings must be dropped. Once the structure of the new process net has been derived, the resource allocation requests of the various places and the labels of its transitions can be readily computed. Incidentally, it is interesting to notice that a straightforward extension of the above computation can also determine the quasi-liveness of the original net $\overline{\mathcal{N}_P}$: This issue can be resolved by checking whether each transition $t \in T$ of $\overline{\mathcal{N}_P}$ constitutes the label of at least one transition in $R(\mathcal{N}_P, M_0)$.

Example. The proposed transformation of the PN-based process representation is exemplified in Figures 5.14 and 5.15. Figure 5.14 depicts the original resource-augmented process net. This net is assumed to have three resource types R_1, R_2 and R_3, with capacities $C_1 = C_2 = C_3 = 12$. For clarity of presentation, the corresponding resource places are not depicted in the figure. Instead, the resource allocation requests posed by the various processing stages are depicted as labels of the corresponding places $p \in P_S$. The STD for the reachable and safe subspace of this net is depicted in Figure 5.15. The total resource allocation taking place at each marking, is annotated next to the corresponding node of the depicted STD. Notice that, for clarity of presentation, both nodes M_{iI} and M_{iF} represent the same marking where the system is idle and empty of any process instances, and they correspond to the process idle place p_0 in the sought PN-based representation.[4] The STD depicted in Figure 5.15 can be readily transformed to a resource-augmented process net that is

---

[4] or, alternatively, marking M_{iI} (resp., marking M_{iF}) corresponds to place i (resp., place o) of Definition 24.

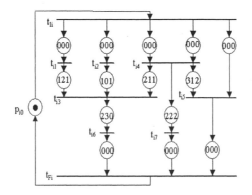

Figure 5.14.   Example: The original resource-augmented process net

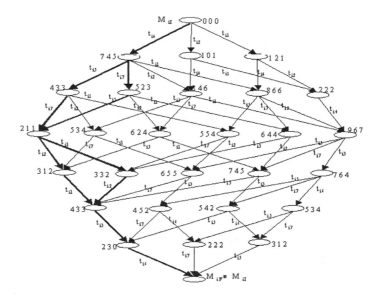

Figure 5.15.   Example: The state machine representing the process reachable and safe subspace

acyclic, quasi-live, strongly reversible, and furthermore, it belongs to the class of DIS-CON-RAS. ◇

The restriction of any single process instance of a process-resource net with acyclic process subnets, $\mathcal{N}$, to the execution sequences encoded by the proposed DIS-CON-RAS-based representation of its reachable and safe subspace, essentially constitutes a supervisory control policy that guarantees the non-blocking execution of that process instance. In fact, this policy constitutes the *maximally permissive supervisor* that guarantees nonblocking execution of a single process instance. It remains to design an additional supervisor that will prevent the development of any deadlock due to the interaction of the various process instances through their competition for the system resources. The availability of the DIS-CON-RAS-based representations for the reachable and safe subspaces of the various resource-augmented process subnets reduces the synthesis of such a supervisor to the development of a liveness-enforcing supervisor for a transformed process-resource net, $\mathcal{N}'$, obtained from $\mathcal{N}$ by replacing its original resource-augmented process subnets with their DIS-CON-RAS-based counterparts. In other words, the broader problem of the LES synthesis for RAS with complex process flows reduces to the simpler problem of synthesizing an appropriate LES for a RAS belonging to the class of DIS-CON-RAS. This problem is addressed in Section 4.

Complexity considerations and the "Thinning" problem. It is important to notice that, in most cases, the reachable and safe subspace of any given resource-augmented process net will possess a super-polynomial size with respect to the size of the original net, and therefore, the problem reduction proposed in this section will not lead to a solution of polynomial complexity with respect to the size of the underlying RAS. The fact that the process reachability spaces are enumerated separately for each resource-augmented process subnet provides considerable alleviation of the underlying computational problems, yet, in many cases, there will be process types involved with a behavioral space that is too large to be handled in a real-time computation. A natural solution to be applied in these cases is to try to *"thin"* the reachable and safe spaces of these processes, by selecting only a subset of their execution sequences to be included in the DIS-CON-RAS model; such a potential selection is depicted by the thicker edges in the STD of Figure 5.15. In particular, bounds can be used to limit the number of execution sequences selected for each process, so that the size of the resulting DIS-CON-RAS model remains polynomial with respect to the size of the original RAS. Then, the challenge is to select these execution sequences in a way that optimizes the performance of the controlled RAS. There are many issues to be considered in the detailed formulation of this problem: The adopted performance criteria, the desired process mix, the applied scheduling policies, the available resource levels, the operation processing times, and other potentially relevant factors can interact so as to make the problem of selecting some 'optimal' set of sequences extremely difficult. We shall refer to the resulting optimization problem as the *"thinning" problem* and we shall provide a canonical characterization for it in Chapter 6, where we consider performance-related issues for logically controlled RAS.

4.   Polynomial-Kernel SCP's for DIS-CON-RAS

This section extends two of the Polynomial-Kernel SCP's developed in Chapter 4 for LIN-SU-RAS, so that they apply to the more complex class of DIS-CON-RAS. As it was shown in the previous section, the availability of these policies is instrumental for the development of PK-LES for the entire class of RAS considered in this book. An additional contribution of this section is that it demonstrates a set of techniques that can be used for the correctness analysis and the optimization of PK-SCP's in the context of the PN-based RAS-modelling framework.

4.1   Implementing Banker's Algorithm for DIS-CON-RAS

Similar to the case of the implementation of Banker's algorithm for the AGV RAS discussed in Section 3 of Chapter 4, a pertinent implementation of Banker's algorithm for the class of DIS-CON-RAS reduces to developing a polynomial procedure for executing Step (2.b) of the algorithm provided in Figure 4.4. Next we outline such a procedure that takes advantage of the particular structure of the process subnets admitted by the class of DIS-CON-RAS. More specifically, since every process subnet, $\mathcal{N}_{P_j}$, that is admitted by the class of DIS-CON-RAS corresponds to strongly connected state machine with every circuit containing its idle place p_{j0}, an active process instance executing some stage Ξ_{jk}, modelled by place $p^* \in P_{S_j}$, can proceed to completion, if and only if there is a path from place p^* to the idle place p_{j0}, such that every transition on it can be executed under the set of resources allocated to that process instance plus the currently free resource units. Such a path can be identified in polynomial computational cost with respect to the size of the underlying resource-augmented process subnet, through a labelling procedure that, starting from the place $p^* \in P_{S_j}$, reaches out and labels all places $p \in P_j$ that satisfy the condition

$$y_r(p) \leq y_r(p^*) + M^+(r), \ \forall r \in P_R \qquad (5.20)$$

In Equation 5.20, $y_r()$ denotes the p-semiflow associated with resource place $r \in P_R$ by Assumption 4, and $M^+(r)$ is the marking of the place $r \in P_R$ that includes all the additional resource units returned by the processes that have already been completed during the previous iterations of the algorithm. The procedure terminates with an accepting response if it manages to label the idle place p_{j0}; otherwise, it rejects.

4.2 Generalized RUN

As indicated by its name, *Generalized (G-) RUN* SCP extends the logic of RUN SCP, that was introduced in Chapter 4 as an algebraic SCP appropriate for LIN-SU-RAS, so that it applies to the broader class of DIS-CON-RAS. This section provides a formal characterization of the policy logic, proves its ability to establish nonblocking operation for the underlying RAS, and discusses a number of techniques that can lead to G-RUN implementations with enhanced operational flexibility. These developments are carried out in the PN-based RAS-modelling framework, in order to take advantage of the results developed in Section 2, but also, in order to familiarize the reader with some typical techniques and arguments for proving SCP correctness in this alternative modelling framework.

Policy Definition

The implementation of G-RUN SCP on any given process-resource net $\mathcal{N} = (P_S \cup P_0 \cup P_R, T, W, M_0)$, modelling a DIS-CON-RAS, involves the specification of two main parameters:

A resource ordering o This concept is similar to the resource ordering involved in the definition of the original RUN SCP, and it is formally defined as a function $o() : \mathcal{R} \rightarrow \{1, \ldots, |\mathcal{R}|\}$. To facilitate the subsequent discussion, we also define, for every place $p \in P_S$, $\rho_p^{min} = \min\{o(R_i) : p \in ||y_{r_i}||, i = 1, \ldots, |\mathcal{R}|\}$, where y_{r_i} denotes the p-semiflow associated with resource R_i according to Assumption 4; in words, ρ_p^{min} denotes the minimum order among the resources supporting the execution of the processing stage corresponding to place p.

A community Ψ over the place set P_S This concept is characterized as follows: Given a place $p \in P_S$, define an *(immediate) neighborhood* of p as a set $N_p \subseteq p^{\bullet\bullet} \cap P_S$. Furthermore, consider a function $g()$, defined on the place set $\{p \in P_S : p^{\bullet\bullet} \cap P_0 = \emptyset\}$, that corresponds to every place p, in its domain, some neighborhood N_p. Then, the community of net $\mathcal{N}$ defined by function $g()$ is the set $\Psi = \{(p, q) : p \in P_S \wedge p^{\bullet\bullet} \cap P_0 = \emptyset \wedge q \in g(p)\}$. The set of all communities Ψ of $\mathcal{N}$ will be denoted by $\mathcal{C_N}$. A community $\Psi \in \mathcal{C_N}$ essentially identifies for every place $p \in P_S$ with $p^{\bullet\bullet} \cap P_0 = \emptyset$, a set of places $q \in P_S$ corresponding to successor stages of p. In the defining logic of G-RUN, the places q with $(p, q) \in \Psi$ constitute the immediate successor stages of p that are guaranteed by the policy to be accessible by every active process instance located in place p. Thus, the parameter Ψ provides a mechanism for managing the process routing flexibility in the policy definition.

G-RUN. Given a resource ordering o and a community $\Psi \in \mathcal{C}_{\mathcal{N}}$, G-RUN is defined as an algebraic SCP $A \cdot M_S \leq f$ s.t. (i) $\dim(f) = |\mathcal{R}|$; (ii) $f_i = C_i$, $i = 1, \ldots, |\mathcal{R}|$; and (iii) the elements α_{ip} of matrix A satisfy the following constraints:

$$\alpha_{ip} \geq \alpha_{iq} \qquad \forall (p,q) \in \Psi, \ \forall R_i \in \mathcal{R} \text{ s.t.} \tag{5.21}$$
$$\exists R_j \in \mathcal{R} \text{ with } o(R_i) \geq o(R_j) \wedge \alpha_{jp} > 0$$
$$C_i \geq \alpha_{ip} \geq y_{r_i}(p) \quad \forall p \in P_S, \ \forall R_i \in \mathcal{R} \tag{5.22}$$
$$\alpha_{ip} \in Z_0^+ \qquad \forall p \in P_S, \ \forall R_i \in \mathcal{R} \tag{5.23}$$

$\diamond$

Similar to the RUN definition for LIN-SU-RAS, G-RUN can be interpreted as a resource reservation scheme. Under this interpretation, Constraint 5.22 requests that every processing stage reserves a number of resource units from each resource R_i that (i) is adequate to satisfy its immediate needs, and (ii) does exceed the resource capacity. Constraint 5.23 asserts the nonnegative integral nature of these reservations. Finally, Constraint 5.21 characterizes the logic underlying the reservation scheme enforced by the policy. In words, Constraint 5.21 requests that for every place $p \in P_S$ with $p^{\bullet\bullet} \cap P_0 = \emptyset$, and every resource R_j with a positive reservation α_{jp} for place p, the resource reservations, α_{ip}, for place p with respect to any resources R_i of order $o(R_i)$ higher than or equal to the order of R_j, are non-decreasing with respect to the corresponding resource reservations for any successor place q such that $(p,q) \in \Psi$. The reader can convince herself that in the operational context of LIN-SU-RAS, this reservation logic encompasses the reservation logic applied by the original RUN SCP.

The Constraint set 5.21-5.23 does not specify uniquely the policy-defining matrix A, even when the employed resource ordering o and the net community Ψ have been fixed. The discussion of Section 2.2, in Chapter 4, suggests that, for a given triplet $< \mathcal{N}, o, \Psi >$, a matrix A that can lead to an efficient G-RUN implementation, can be obtained as an optimal solution to the following integer program:

$$\min G(A; \mathcal{N}, o, \Psi) = \sum_{i=1}^{|\mathcal{R}|} \sum_{p \in P_S} \alpha_{ip} \tag{5.24}$$

s.t.

Equation 5.21 - 5.23

In fact, the integrality requirement for α_{ip}, posed by Constraint 5.23, need not be explicitly introduced in the above formulation, since Constraint 5.21 possesses a network structure, which when combined with the integrality of C_i, for all i, implies that the polyhedron defined by Constraints 5.21-5.23 will have

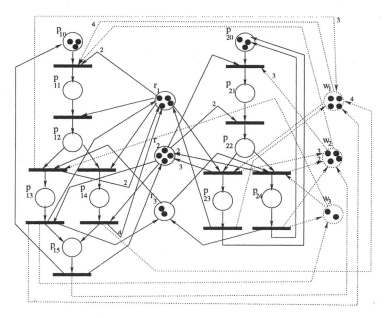

Figure 5.16. Example: G-RUN implementation on a process-resource net

integral extreme points. In addition, a natural way to satisfy Constraint 5.21 while remaining consistent with the objective of Equation 5.24, is to set $\alpha_{pi} = 0$, $\forall R_i \in \mathcal{R}$ s.t. $o(R_i) < \rho_p^{min}$. Hence, a simpler formulation for determining the A matrix during a G-RUN implementation, is as follows:

$$\min G(A; \mathcal{N}, o, \Psi) = \sum_{i=1}^{|\mathcal{R}|} \sum_{p \in P_S} \alpha_{ip} \qquad (5.25)$$

s.t.

$$\alpha_{ip} \geq \alpha_{iq} \qquad \forall (p,q) \in \Psi, \forall R_i \in \mathcal{R} \text{ with } o(R_i) \geq \rho_p^{min} \ (5.26)$$

$$\alpha_{ip} = 0 \qquad \forall p \in P_S, \forall R_i \in \mathcal{R} \text{ with } o(R_i) < \rho_p^{min} \qquad (5.27)$$

$$C_i \geq \alpha_{ip} \geq y_{r_i}(p) \quad \forall p \in P_S, \forall R_i \in \mathcal{R} \qquad (5.28)$$

The formulation of Equations 5.25-5.28 is a linear program polynomially sized, in terms of variables and constraints, with respect to the underlying process-resource net, and therefore, the complexity of obtaining the corresponding G-RUN implementation is of polynomial computational cost with respect to the size of the underlying RAS. The next example demonstrates the implementation of G-RUN SCP on a given process-resource net modelling a DIS-CON-RAS.

Example. Consider the DIS-CON-RAS depicted in Figure 5.16, consisting of three resource types R_1, R_2 and R_3, with capacities $C_1 = C_2 = 4, C_3 = 2$, and supporting two process types Π_1 and Π_2. Process type Π_1 is defined by the set of partially ordered processing stages corresponding to places p_{11}, p_{12}, p_{13}, p_{14} and p_{15}, while process type Π_2 is defined by the processing stages corresponding to places p_{21}, p_{22}, p_{23} and p_{24}. The resource allocation requests associated with each of these processing stages are provided by the net p-semiflows y_{r_i}, $i = 1, 2, 3$, characterizing the reusability of the system resources; specifically, under the place ordering $< p_{10}, p_{11}, p_{12}, p_{13}, p_{14}, p_{15}, p_{20}, p_{21}, p_{22}, p_{23}, p_{24}, r_1, r_2, r_3$ $>$, these p-semiflows are as follows: $y_{r_1} = (0, 2, 3, 1, 4, 0, 0, 0, 0, 1, 0, 1, 1, 1)$, $y_{r_2} = (0, 0, 0, 0, 0, 1, 0, 1, 3, 0, 1, 1, 1, 1)$, $y_{r_3} = (0, 0, 0, 1, 0, 0, 0, 0, 0, 0, 1, 1, 1, 1)$. Applying the LP formulation of Equations 5.25-5.28, under the arbitrarily selected resource ordering $o(R_1) = 2$, $o(R_2) = 3$, $o(R_3) = 1$, and the community $\Psi = \{(p_{11}, p_{12}), (p_{12}, p_{14}), (p_{13}, p_{15}), (p_{14}, p_{15}), (p_{21}, p_{22}), (p_{22}, p_{23})\}$, we obtain the G-RUN implementation expressed by the following constraints

$$\begin{pmatrix} 4 & 4 & 1 & 4 & 0 & 0 & 0 & 1 & 0 \\ 1 & 1 & 1 & 1 & 1 & 3 & 3 & 0 & 1 \\ 0 & 0 & 1 & 0 & 0 & 0 & 0 & 0 & 1 \end{pmatrix} \cdot M_S \leq \begin{pmatrix} 4 \\ 4 \\ 2 \end{pmatrix} \qquad (5.29)$$

The control subnet enforcing the above constraint set on the original process-resource net can be generated according to the theory of control-place invariants, discussed in Section 3.1, and it is depicted in Figure 5.16 with dotted lines. $\diamond$

Establishing the correctness of G-RUN SCP

An immediate implication of the discussion of Section 3.1 is that the PN-based implementation of G-RUN SCP on any process-resource net modelling a DIS-CON-RAS, can be represented by another process-resource net that remains in the class of DIS-CON-RAS. Furthermore, the state-machine structure of the involved process subnets, combined with Constraint 5.22, imply that the controlled net preserves the quasi-liveness of the underlying process subnets. In the light of these remarks, the correctness of G-RUN SCP – or equivalently, the liveness and reversibility of the resulting controlled net – can be established by showing that this net satisfies the condition of Theorem 5.3. This result is formally stated and proved in the following theorem.

THEOREM 5.7 *The controlled net $\mathcal{N}^c = (P_S \cup P_0 \cup P_R \cup P_W, T, W^c, M_0^c)$, corresponding to a G-RUN implementation on a process-resource net $\mathcal{N} = (P_S \cup P_0 \cup P_R, T, W, M_0)$ belonging to the class of DIS-CON-RAS, has no resource-induced deadly marked siphons in its modified reachability space $R(\mathcal{N}^c, M_0^c)$.*

Proof: We prove this result by contradiction. So, let $A = [\alpha_{ip}]_{p \in P_S}^{i=1,...,|\mathcal{R}|}$ denote the policy-defining matrix for the considered G-RUN implementation, and $\Psi \in \mathcal{C}_{\mathcal{N}}$ denote a community for which A satisfies Constraints 5.21-5.23 of the G-RUN definition. Furthermore, suppose that there exists a modified marking $\overline{M'} \in \overline{R(\mathcal{N}^c, M_0^c)}$ that contains a deadly marked siphon, S, s.t. $S \cap (P_R \cup P_W) \neq \emptyset$ and every place $p \in S \cap (P_R \cup P_W)$ disables some transition of $\mathcal{N}^c$ in $\overline{M'}$. Then, by applying an argument similar to that in the proof of Lemma 7, we can establish the existence of a marking $M \in R(\mathcal{N}^c, M')$, s.t. (i) the modified marking $\overline{M}$ contains the resource-induced deadly marked siphon S; (ii) $\{p \in P_S : \overline{M}(p) \geq 1\} \neq \emptyset$; and (iii) for all places $p \in P_S$ with $\overline{M}(p) \geq 1$, every transition $t \in p^\bullet$ is dead at $R(\mathcal{N}^c, M)$.

Since S is a resource-induced deadly marked siphon at $\overline{M}$, the place set $(S_R \cup S_W)$ with $S_R = \{p \in P_R : \overline{M}(p) < M_0(p)\}$ and $S_W = \{p \in P_W : \overline{M}(p) < M_0(p)\}$, is non-empty. Consider a place $q_1 \in S_R \cup S_W$. From the above definitions it follows that there exists a place $p_1 \in P_S$ s.t. $M(p_1) \geq 1$, $p_1 \in ||y_{q_1}^c||$,[5] and every transition $t \in p_1^\bullet$ is dead. Select a transition $t_1 \in p_1^\bullet$ s.t. $(p_1, \eta_1 \equiv t_1^\bullet \cap P_S) \in \Psi$. Since t_1 is dead, $M(p_1) \geq 1$, and the involved process subnets are state machines, there exists a place $q_2 \in S_R \cup S_W$ disabling t_1. Repeating the above argument on place q_2, and considering the finiteness of set $S_R \cup S_W$, we infer the existence of a set $\{q_1, q_2, \ldots, q_k\} \subseteq S_R \cup S_W$, and a corresponding set $\{p_1, p_2, \ldots, p_k\} \subseteq P_S$ s.t. p_i, $i = 1, \ldots, k$, is a marked place in M belonging to $||y_{q_i}^c||$. Furthermore, there exists some transition $t_i \in p_i^\bullet$ that is disabled by place q_{i+1}, for $i = 1, \ldots, k-1$, and a transition $t_k \in p_k^\bullet$ that is disabled by some place q_i with $i \in \{1, \ldots, k\}$.

Next, consider a place $q_{i^*} \in \{q_1, q_2, \ldots, q_k\}$ with $i^* = \arg\min_i\{o(q_i)\}$, where $o(q_i)$ is the partial ordering employed in the policy implementation, extended to set $P_R \cup P_W$ by assigning the same order to each resource r_l and its corresponding control place w_l. Also, let p_{i^*} and η_{i^*} respectively denote the places in $P_S \cap ||y_{q_{i^*}}^c||$ and in $P_S \cap p_{i^*}^{\bullet\bullet}$, that are induced by place q_{i^*} in the cyclic construction of the previous paragraph. Finally, for notational convenience, let u (resp., v) denote the index of the resource or control place q_{i^*} (resp., q_{i^*+1}) in the corresponding place set. Then, $p_{i^*} \in ||y_{q_{i^*}}^c||$ implies that $\alpha_{up_{i^*}} > 0$. In the same manner, $\alpha_{v\eta_{i^*}} > 0$. Since the pair $(p_{i^*}, \eta_{i^*}) \in \Psi$ satisfies Constraint 5.21 and $o(R_u) \leq o(R_v)$, it must be that $\alpha_{vp_{i^*}} \geq \alpha_{v\eta_{i^*}} > 0$. Hence, $p_{i^*} \in ||y_{w_v}^c||$, and in the control subnet representing the considered G-RUN implementation, $W^c(w_v, t_{i^*}) = 0$. This last remark further implies that q_{i^*+1} corresponds to a resource and not to a control place. Therefore, $\alpha_{vp_{i^*}} \geq \alpha_{v\eta_{i^*}} \geq y_{r_v}^c(\eta_{i^*}) = y_{r_v}^c(p_{i^*}) + W^c(r_v, t_{i^*})$. Hence, $y_{w_v}^c(p_{i^*}) - y_{r_v}^c(p_{i^*}) \geq W^c(r_v, t_{i^*})$. Since place r_v disables transition t_{i^*} at M, it must be $0 \leq M(r_v) < W^c(r_v, t_{i^*})$. These inequalities, combined with the facts that $M(w_v) \geq 0$, $M(p_{i^*}) \geq 1$,

---

[5] $y_{q_1}^c$ denotes the p-semiflow associated with place q_1 in the controlled net $\mathcal{N}^c$, through Assumption 4.

$||y_{w_v}^c|| \supseteq ||y_{r_v}^c||$, and $y_{w_v}^c(p) \geq y_{r_v}^c(p)$, $\forall p \in P_S$, imply that

$$\sum_{p \in ||y_{w_v}^c||} y_{w_v}^c(p) M(p) - \sum_{p \in ||y_{r_v}^c||} y_{r_v}^c(p) M(p) \geq$$

$$(y_{w_v}^c(p_{i^*}) - y_{r_v}^c(p_{i^*})) M(p_{i^*}) + M(w_v) - M(r_v) > 0$$

The last set of inequalities contradicts the fact that

$$\sum_{p \in ||y_{w_v}^c||} y_{w_v}^c(p) M(p) = \sum_{p \in ||y_{r_v}^c||} y_{r_v}^c(p) M(p) = C_v$$

and proves the theorem. ◇

Enhancing the policy permissiveness

In this section we consider three techniques that, when applied during the implementation of G-RUN SCP on any given DIS-CON-RAS, can lead to a policy instantiation with enhanced operational flexibility. The first two of these techniques are essentially a detailed implementation, in the context of the G-RUN SCP, of the ideas discussed in Section 2.2 of Chapter 4, regarding an optimized parametrization of RUN SCP and the possibility of disjunctive implementations obtained through different parameterizations of the policy. The third technique capitalizes on the liveness verification test developed in Section 2.2, in order to systematically relax the right-hand-side vector of any G-RUN implementation, while maintaining its correctness.

"Optimal" parameter selection for G-RUN implementations. As it was discussed in Section 2.2 of Chapter 4, a pertinent selection of the G-RUN parameters of the resource ordering o and the community Ψ, during the policy implementation on any given DIS-CON-RAS configuration, should seek to minimize the total magnitude of the reservations enforced by the policy. This minimization can be carried out by solving the following Mixed Integer Programming formulation:

$$\min \sum_{i=1}^{|\mathcal{R}|} \sum_{p \in P_S} X_{ip} \tag{5.30}$$

s.t.

$$\sum_{i=1}^{|\mathcal{R}|} Y_i = \frac{|\mathcal{R}|^2 + |\mathcal{R}|}{2} \tag{5.31}$$

$$Y_i \geq 1 \qquad\qquad \forall R_i \in \mathcal{R} \tag{5.32}$$

$$Y_i \geq Y_p^{min} \qquad \forall p \in P_S, \forall R_i \in Q_p \tag{5.33}$$

$P_{NT} \equiv \{p \in P_S : p^{\bullet\bullet} \cap P_0 = \emptyset\}$

$F_p \equiv \{q \in P_S : q \in p^{\bullet\bullet}\}, \forall p \in P_S$

$Q_p \equiv \{R_i \in \mathcal{R} : p \in ||y_{r_i}||\}$

μ: big-M parameter set to $\max\{C_i : i = 1, \ldots, |\mathcal{R}|\}$

ν: big-M parameter set to $|\mathcal{R}| - 1$

$X_{ip} \equiv$ nonnegative real-valued variable returning the value of α_{ip} in the optimized G-RUN matrix A, $\forall R_i \in \mathcal{R}$, $\forall p \in P_S$

$Y_i \equiv$ a positive integer-valued variable returning the order $o(R_i)$ for the optimized G-RUN implementation, $\forall R_i \in \mathcal{R}$

$Y_p^{min} \equiv \rho_p^{min}$ in the optimized G-RUN implementation, $\forall p \in P_S$

$Z_{ip} \equiv I_{\{Y_i = Y_p^{min}\}}, \forall p \in P_S, \forall R_i \in Q_p$

$\Lambda_{ip} \equiv I_{\{Y_i \geq Y_p^{min}\}}, \forall R_i \in \mathcal{R}, \forall p \in P_{NT}$

$\Gamma_{pq} \equiv I_{\{(p,q) \in \Psi\}}$, where Ψ is the community employed by the optimized G-RUN implementation, $\forall p \in P_{NT}, \forall q \in F_p$

$E_{ipq}^1 \equiv I_{\{\Lambda_{ip} = 1 \wedge \Gamma_{pq} = 1\}}, \forall R_i \in \mathcal{R}, \forall p \in P_{NT}, \forall q \in F_p$

$E_{ipq}^2 \equiv I_{\{\Lambda_{ip} = 0 \wedge \Gamma_{pq} = 1\}}, \forall R_i \in \mathcal{R}, \forall p \in P_{NT}, \forall q \in F_p$

Figure 5.17. Definition of the various sets, parameters and variables involved in the formulation of Equations 5.30 - 5.49

$$\sum_{R_i \in Q_p} Z_{ip} \geq 1 \qquad \forall p \in P_S \qquad (5.34)$$

$$\nu(1 - Z_{ip}) \geq Y_i - Y_p^{min} \qquad \forall p \in P_S, \forall R_i \in Q_p \qquad (5.35)$$

$$(\nu + 1)\Lambda_{ip} \geq Y_i - Y_p^{min} + 1 \qquad \forall p \in P_{NT}, \forall R_i \in \mathcal{R} \qquad (5.36)$$

$$\nu(1 - \Lambda_{ip}) \geq Y_p^{min} - Y_i \qquad \forall p \in P_{NT}, \forall R_i \in \mathcal{R} \qquad (5.37)$$

$$\sum_{q \in F_p} \Gamma_{pq} \geq 1 \qquad \forall p \in P_{NT} \qquad (5.38)$$

$$\Gamma_{pq} \geq E_{ipq}^1 \quad \forall p \in P_{NT}, \forall q \in F_p, \forall R_i \in \mathcal{R} \quad (5.39)$$

$$\Lambda_{ip} \geq E_{ipq}^1 \quad \forall p \in P_{NT}, \forall q \in F_p, \forall R_i \in \mathcal{R} \quad (5.40)$$

$$E_{ipq}^1 + 1 \geq \Gamma_{pq} + \Lambda_{ip} \quad \forall p \in P_{NT}, \forall q \in F_p, \forall R_i \in \mathcal{R} \quad (5.41)$$

$$\Gamma_{pq} \geq E_{ipq}^2 \quad \forall p \in P_{NT}, \forall q \in F_p, \forall R_i \in \mathcal{R} \quad (5.42)$$

$$1 - \Lambda_{ip} \geq E_{ipq}^2 \quad \forall p \in P_{NT}, \forall q \in F_p, \forall R_i \in \mathcal{R} \quad (5.43)$$

$$E_{ipq}^2 \geq \Gamma_{pq} - \Lambda_{ip} \quad \forall p \in P_{NT}, \forall q \in F_p, \forall R_i \in \mathcal{R} \quad (5.44)$$

$$X_{iq} - X_{ip} \leq \mu(1 - E_{ipq}^1) \quad \forall p \in P_{NT}, \forall q \in F_p, \forall R_i \in \mathcal{R} \quad (5.45)$$

$$X_{ip} \leq \mu(1 - E_{ipq}^2) \quad \forall p \in P_{NT}, \forall q \in F_p, \forall R_i \in \mathcal{R} \quad (5.46)$$

$$C_i \geq X_{ip} \geq y_{r_i}(p) \qquad \forall R_i \in \mathcal{R}, \forall p \in P_S \qquad (5.47)$$

$$Y_i \in \mathbb{Z}^+ \qquad \forall R_i \in \mathcal{R} \qquad (5.48)$$

$$Z_{ip}, \Lambda_{ip}, \Gamma_{pq}, E^1_{ipq}, E^2_{ipq} \in \{0,1\} \quad \forall p \in P_S, \; \forall q \in P_S, \; \forall R_i \in \mathcal{R} \quad (5.49)$$

The definition of the various sets, parameters and variables that are necessary for the understanding of this formulation are provided in Figure 5.17. To further assist with the understanding of this formulation, we notice that Constraints 5.31 and 5.32 essentially define Y_i as a resource ordering, while recognizing that the search for an optimal resource ordering can be confined among the *total* orderings of this set. Constraints 5.33 to 5.35 seek to enforce the definition of variables Y_p^{min}. Similarly, Constraints 5.36 and 5.37 enforce the definition of variables Λ_{ip}, while Constraint 5.38 guarantees that the selected community Ψ contains a successor stage for every non-terminal place $p \in P_{NT}$. Constraints 5.39 to 5.41 enforce the definition of variables E^1_{ipq}, while Constraints 5.42 to 5.44 enforce the definition of variables E^2_{ipq}. Finally, Constraints 5.45 to 5.47 correspond to the definition of the G-RUN logic, itself.

Example. Applying the MIP formulation of Equations 5.30-5.49 to the DIS-CON-RAS of Figure 5.16, we obtained the optimized resource ordering $o^*(R_1) = 3, o^*(R_2) = 2, o^*(R_3) = 1$, and the optimized community $\Psi^* = \{(p_{11}, p_{12}), (p_{12}, p_{13}), (p_{13}, p_{15}), (p_{14}, p_{15}), (p_{21}, p_{22}), (p_{22}, p_{24})\}$. The resulting G-RUN implementation is given by the following set of inequalities:

$$\begin{pmatrix} 3 & 3 & 1 & 4 & 0 & 0 & 0 & 1 & 0 \\ 0 & 0 & 1 & 0 & 1 & 3 & 3 & 0 & 1 \\ 0 & 0 & 1 & 0 & 0 & 0 & 0 & 0 & 1 \end{pmatrix} \cdot M_S \leq \begin{pmatrix} 4 \\ 4 \\ 2 \end{pmatrix} \quad (5.50)$$

The reader can verify that the policy-defining matrix in the G-RUN implementation of Equation 5.50 is element-wise smaller than the policy-defining matrix in the G-RUN implementation of Equation 5.29. Hence, the G-RUN implementation of Equation 5.50 will tend to enable a higher degree of concurrency than the G-RUN implementation of Equation 5.29. On the other hand, it must be noticed that some routing options that were guaranteed by the G-RUN implementation of Equation 5.29 might not be guaranteed by the G-RUN implementation of Equation 5.50, since these policies engage different communities in their definition.

"Orthogonal" disjunctive G-RUN implementations. We remind the reader that the basic motivation underlying the concept of disjunctive PK-SCP implementations is the observation that the resulting policy admits the union of the subspaces admitted by the constituent policies. Furthermore, the notion of "orthogonal" parameterizations that are to be employed during a disjunctive implementation of RUN SCP, seeks to systematically increase the complementarity of the constituent policy instantiations, by ensuring that each such new instantiation corresponds to a resource reservation scheme that is differentiated as much as possible from the reservations schemes corresponding to the already selected policy instantiations. This idea is operationalized for the more general case of G-RUN SCP, as follows: Given a process-resource net $\mathcal{N}$, and a set of G-RUN instantiations, Δ_j, $j = 1, \ldots, l$, for it, with corresponding policy-defining matrices A^j, let $\overline{A}^l$ be the matrix with elements $\overline{a}_{ip}^l = \max_{j=1}^l \{a_{ip}^j\}$, if $l \geq 1$; 0, otherwise. Matrix $\overline{A}^l$ indicates the stage-resource pairs that present the most stringent reservations in the scope of the already selected policy instantiations. Therefore, the matrix $A^{(l+1)}$ to be employed by the next policy instantiation, Δ_{l+1}, is obtained by solving the following formulation:

$$\min \sum_{i=1}^{|\mathcal{R}|} \sum_{p \in P_S} (\overline{a}_{ip}^l + \epsilon) X_{ip} \tag{5.51}$$

s.t.

Equations 5.31 to 5.49

The constraints of the above formulation are identical to the constraints employed by the formulation that computes the "optimal" parameter selection for a single G-RUN instantiation, since in both cases, they seek to enforce the requirement that the returned matrix $[X_{ip}]$ satisfies the conditions listed in the G-RUN definition. The term $\overline{a}_{ip}^l$ in the cost coefficients of the objective function of Equation 5.51 implements the notion of "orthogonality" discussed above, since the "inner-product" $\sum_{i=1}^{|\mathcal{R}|} \sum_{p \in P_S} \overline{a}_{ip}^l X_{ip}$, of matrix $\overline{A}^l$ with the matrix $A^{(l+1)}$, will tend to be minimized by matching the largest values of X_{ip} with the smallest values of $\overline{a}_{ip}^l$ and vice versa. The second cost term, ϵ, should be taken such that $0 < \epsilon \ll 1$, and it intends to penalize unnecessarily large reservations incurred by the new policy Δ_{l+1}, especially for elements X_{ip} corresponding to $\overline{a}_{ip}^l = 0$. The complete algorithm for selecting (up to) k instantiations for a disjunctive G-RUN implementation according to the idea of "orthogonal" policy parameterizations, is given in Figure 5.18. Notice that, as expected, the first iteration in the algorithm of Figure 5.18 corresponds to the solution of the "optimal" parameter selection problem for a single G-RUN instantiation.

Example. We demonstrate the concept of an "orthogonal" disjunctive G-RUN implementation, by applying it on the DIS-CON-RAS of Figure 5.16, with $k = 2$ and $\epsilon = 0.1$. Based on the previous discussion, A^1 is given by Equation 5.50. For $l = 2$, we solve the MIP formulation of Equations 5.51, 5.31 - 5.49 with $\overline{A}^1 = A^1$. The new resource ordering and community are respectively $o^2(R_1) = 3$, $o^2(R_2) = 1$, $o^2(R_3) = 2$, and $\Psi^2 = \{(p_{11}, p_{12}), (p_{12}, p_{13}), (p_{13}, p_{15}), (p_{14}, p_{15}), (p_{21}, p_{22}), (p_{22}, p_{24})\}$. The new policy-defining constraints

1 $\overline{A} = 0; \; l = 1$

2 **while** $l \leq k$ **do**

   (a) Solve the MIP of Equations 5.51, 5.31 - 5.49, to obtain the new policy matrix A^l.

   (b) $\overline{A}' := A \vee A^l$, where $\vee$ denotes the element-wise max operator.

   (c) **If** $\overline{A}' \neq \overline{A}$ **then** $\overline{A} := \overline{A}'; \; l := l + 1;$ **else exit;**

   endwhile

Figure 5.18.   "Orthogonal" k-disjunctive G-RUN implementation

are:

$$\begin{pmatrix} 3 & 3 & 1 & 4 & 0 & 0 & 0 & 1 & 0 \\ 0 & 0 & 0 & 0 & 1 & 3 & 3 & 0 & 1 \\ 0 & 0 & 1 & 0 & 0 & 1 & 1 & 0 & 1 \end{pmatrix} \cdot M_S \leq \begin{pmatrix} 4 \\ 4 \\ 2 \end{pmatrix} \qquad (5.52)$$

We also notice, for completeness, that $\overline{A}^2$ is defined by:

$$\overline{A}^2 \equiv \overline{A}^1 \vee A^1 = \begin{pmatrix} 3 & 3 & 1 & 4 & 0 & 0 & 0 & 1 & 0 \\ 0 & 0 & 1 & 0 & 1 & 3 & 3 & 0 & 1 \\ 0 & 0 & 1 & 0 & 0 & 1 & 1 & 0 & 1 \end{pmatrix} \qquad (5.53)$$

G-RUN constraint relaxation by means of the liveness criterion of Corollary 5.   Another way to enhance the permissiveness of a G-RUN implementation on a given DIS-CON-RAS, is by trying to relax the right-hand-side (rhs) vector of the policy-defining constraints, while maintaining the policy correctness. More specifically, given a process-resource net $\mathcal{N} = (P_S \cap P_0 \cap P_R \cap P_W, T, W, M_0)$, modelling a DIS-CON-RAS controlled by a G-RUN implementation expressed by the inequalities $A \cdot M_S \leq f_0 \equiv C$, one can seek to identify vectors $f \geq f_0$ that will maintain the liveness of the controlled net, when replacing f_0 in the policy-defining constraints. We remind the reader that in the PN-based representation of the policy constraints, the rhs-vector f specifies the initial marking of the control places $w_i \in P_W$ (c.f. Equation 5.17). Hence, one can integrate easily the liveness criterion of Corollary 5 in a search procedure that seeks to identify maximal vectors f leading to correct policy implementations. This search must be conducted over a lattice $\{f \in (Z^+)^{|\mathcal{R}|} : f_0 \leq f \leq \overline{f}\}$, where $\overline{f}$ is determined by the maximal process concurrency that can be supported by the resource availability. A practical way to compute the components of $\overline{f}$ is by solving the following linear programs:

$$\forall i \in \{1,\ldots,\mathcal{R}\},\ \overline{f}(i) =$$

$$\max_{\{x_p:p\in P_S \wedge \alpha_{ip}>0\}} \sum_{\{p\in P_S:\alpha_{ip}>0\}} \alpha_{ip}x_p$$

s.t.

$$\sum_{\{p\in P_S:\alpha_{ip}>0\}} y_{r_i}(p)x_p \leq C_i,\ \forall i \in \{1,\ldots,\mathcal{R}\}$$

$$x_p \geq 0,\ \forall p \in P_S \text{ with } \alpha_{ip} > 0 \tag{5.54}$$

In the formulation of Equation 5.54, α_{ip} denote the elements of the policy-defining matrix A, and y_{r_i} is the p-semiflow associated with resource R_i by Assumption 4. In the general case, the correctness of the G-RUN implementation resulting from a vector f in the aforementioned lattice does not depend monotonically on the value of f. Therefore, a systematic search for maximal elements f^* resulting to correct policy instantiations must start from the largest vector $\overline{f}$, generate the arborescence defined by the "$\leq$" order, and terminate when an element f^* has been identified on every path of this arborescence.

Example. Consider the G-RUN implementation for the DIS-CON-RAS of Figure 5.16 that is expressed by Equation 5.29. Application of the LP formulation of Equation 5.54 to this policy implementation returns the vector $\overline{f} = (8,15,2)^T$. Subsequent application of the search algorithm outlined above on the lattice $\{f \in (Z^+)^{|\mathcal{R}|} : (4,4,2)^T \leq f \leq (8,15,2)^T\}$ results in the unique maximal element $f^* = (7,8,2)^T$. Hence, a correct relaxed policy implementation is defined by the following constraint set:

$$\begin{pmatrix} 4 & 4 & 1 & 4 & 0 & 0 & 0 & 1 & 0 \\ 1 & 1 & 1 & 1 & 1 & 3 & 3 & 0 & 1 \\ 0 & 0 & 1 & 0 & 0 & 0 & 0 & 0 & 1 \end{pmatrix} \cdot M_S \leq \begin{pmatrix} 7 \\ 8 \\ 2 \end{pmatrix} \tag{5.55}$$

Clearly, this new policy implementation admits all the RAS states admitted by the original policy of Equation 5.29. Furthermore, the reader can check that this new policy admits, for instance, the RAS state expressed by $M_S = (0,1,0,0,0,1,1,0,0)^T$ that is not admitted by the original policy implementation, i.e., the attempted relaxation of the policy constraints is effective. $\diamond$

We conclude this section by noticing that the techniques discussed above can be applied in a complementary and synergistic fashion; in particular, the third technique, aiming at the relaxation of the rhs-vector of the policy-defining constraints, can be applied to any G-RUN implementation(s) derived by the first two approaches. An additional line of research aiming at the development of more efficient G-RUN implementations, has sought to further extend the idea of disjunctive G-RUN implementation, to that of a *"policy mixture"*, where the various executing process instances are *dynamically* (re-)distributed to a number of G-RUN instantiations, in an effort to balance the underlying resource reservations; we refer the reader to (Park and Reveliotis, 2002b) for a more extensive treatment of this idea.

5.   Polynomial-Kernel SCP's for RAS with Uncontrollable Behavior

This section revisits the problem of nonblocking supervision for RAS presenting Type-1 and/or Type-2 uncontrollability, seeking to develop variations of the G-RUN SCP that are appropriate for each of these cases. More specifically, the first part of this section extends the G-RUN-SCP so that it applies on DIS-CON-RAS with Type-1 uncontrollability. The second part of the section extends the DIS-CON-RAS class so that it can model the behavior resulting from Type-2 uncontrollability, and subsequently it establishes that the G-RUN SCP can provide effective liveness-enforcing supervision for the extended RAS class. The last part of the section outlines the extension of these results to broader RAS classes through the problem reduction technique introduced in Section 3.2.

5.1   G-RUN implementation for DIS-CON-RAS with Type-1 Uncontrollability

Type-1 uncontrollability is modelled in the PN-based modelling framework for DIS-CON-RAS, by partitioning the set of transitions, T, to two subsets, T_c and T_u, denoting respectively the process advancement events that correspond to controllable and uncontrollable resource allocations. A liveness-enforcing supervisor is considered to be *acceptable* for the resulting class of nets, in the spirit of Definition 13, *iff* it does not seek to disable any transition $t \in T_u$. In the particular case of algebraic PK-SCP's, the notion of acceptability can be further elaborated by taking advantage of the theory of control-place invariants, that governs the implementation of these policies in the PN-based modelling framework. More specifically, Equation 5.16 of Section 3.1 implies the following acceptability test.

PROPOSITION 4 *Given a process-resource net* $\mathcal{N} = (P_S \cup P_0 \cup P_R, T_c \cup T_u, W, M_0)$, *modelling a DIS-CON-RAS with Type-1 uncontrollability, let* Θ_u *denote the part of the net flow matrix* Θ *corresponding to the uncontrollable transitions* $t \in T_u$. *Then, any given algebraic SCP,* $A \cdot M_S \leq b$, *will be an acceptable SCP for* $\mathcal{N}$, *iff*

$$A_{[l,\cdot]} \cdot \Theta_u \leq 0, \ \forall l \tag{5.56}$$

Proof: From the definition of the flow matrix Θ, for any place p and transition t of a pure PN $\mathcal{N}$, $W(p,t) = 0$ *iff* $\Theta(p,t) \geq 0$. But then, Equation 5.16 implies that, for any control place $w_l \in P_W$ and transition $t \in T_u$, $W(w_l,t) = 0$ *iff* $A_{[l,\cdot]} \cdot \Theta_{[\cdot,t]} \leq 0$. The proof concludes by noticing that Equation 5.56 is a collective statement of this result over all control places $w_l \in P_W$ and transitions $t \in T_u$. $\diamond$

Proposition 4 provides the necessary guideline for the development of acceptable G-RUN implementations for DIS-CON-RAS with Type-1 uncontrollability. The key idea is to substitute the set of Constraints 5.26-5.28 that were employed in the policy implementations on totally controllable DIS-CON-RAS, with an alternative set of constraints that satisfy (i) the original policy-defining conditions, expressed by Equations 5.21-5.23, and also (ii) the condition of Equation 5.56. Such a G-RUN implementation is provided in the next definition.

U-G-RUN. Given a process-resource net, $\mathcal{N} = (P_0 \cup P_S \cup P_R, T_c \cup T_u, W, M_0)$, modelling a DIS-CON-RAS with Type-1 uncontrollability, the algebraic LES $A \cdot M_S \leq f$ is a U-G-RUN LES *iff* (i) $\dim(f) = |\mathcal{R}|$; (ii) $f_i = C_i$, $i = 1, \ldots, |\mathcal{R}|$; and (iii) $\exists\, o() : \mathcal{R} \to \{1, \ldots, |\mathcal{R}|\}$ and some community $\Psi \in \mathcal{C}_\mathcal{N}$, s.t.

$$\alpha_{ip} \geq \alpha_{iq} \quad \forall (p,q) \in \Psi, \forall R_i \in \mathcal{R}, \text{ s.t.} \tag{5.57}$$
$$p^\bullet \cap {}^\bullet q \in T_c \wedge o_i \geq \rho_p^{min}$$

$$\alpha_{ip} = 0 \quad \forall (p,q) \in \Psi, \forall R_i \in \mathcal{R}, \text{ s.t.} \tag{5.58}$$
$$p^\bullet \cap {}^\bullet q \in T_c \wedge o_i < \rho_p^{min}$$

$$\alpha_{ip'} > \alpha_{iq'} \quad \forall (p',q'), \forall R_i \in \mathcal{R}, \text{ s.t.} \tag{5.59}$$
$$p', q' \in P_S \wedge p'^\bullet \cap {}^\bullet q' \in T_u$$

$$C_i \geq \alpha_{ip} \geq y_{r_i}(p) \quad \forall p \in P_S, \forall R_i \in \mathcal{R} \tag{5.60}$$

$$\alpha_{ip} \in Z_0^+ \quad \forall p \in P_S, \forall R_i \in \mathcal{R} \tag{5.61}$$

In the above set of equations, ρ_p^{min} and $y_{r_i}(p)$ should be interpreted as in the prior definitions of G-RUN SCP. The next theorem establishes that the conditions expressed by Equations 5.57-5.61 satisfy the original definition of G-RUN SCP while the resulting policy implementation is acceptable with respect to the uncontrollability of the underlying RAS.

THEOREM 5.8 *Consider a process-resource net, $\mathcal{N} = (P_0 \cup P_S \cup P_R, T_c \cup T_u, W, M_0)$, modelling a DIS-CON-RAS with Type-1 uncontrollability, and a U-G-RUN realization on it. Then, (i) the considered supervisor satisfies the G-RUN-defining conditions, expressed by Equations 5.21-5.23, and (ii) it is acceptable with respect to the net uncontrollability.*

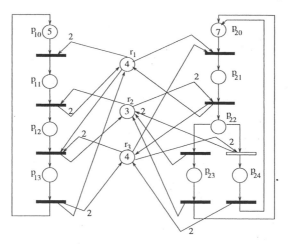

Figure 5.19. Example: The process-resource net modelling the considered DIS-CON-RAS

Proof: Part (i) of Theorem 5.8 results immediately from the following two observations: For all $(p, q) \in \Psi$ with $p^\bullet \cap {}^\bullet q \in T_c$, the conditions expressed by Equations 5.57-5.60 are identical to the conditions expressed by Equations 5.26-5.28. For $(p, q) \in \Psi$ with $p^\bullet \cap {}^\bullet q \in T_u$, the condition of Equation 5.59 is stronger than the condition expressed by Equation 5.21 in the original definition of G-RUN.

To prove part (ii) of Theorem 5.8, it suffices to show that $A_{[i,\cdot]} \cdot \Theta_{[\cdot,t]} \leq 0$, for all $i \in \{1, \ldots, |\mathcal{R}|\}$ and $t \in T_u$. Hence, consider an uncontrollable transition $t \in T_u$, and its column $\Theta_{[\cdot,t]}$ in the flow matrix of $\mathcal{N}$. Let $p' = {}^\bullet t \cap P_S$ and $q' = t^\bullet \cap P_S$. Since $\mathcal{N}$ is a process-resource net corresponding to DIS-CON-RAS, $\Theta_{[\cdot,t]}(p') = -1$, $\Theta_{[\cdot,t]}(q') = 1$, and $\Theta_{[\cdot,t]}(p) = 0$, $\forall p \in P_S \backslash \{p', q'\}$. Then, the fact that $A_{[i,\cdot]} \cdot \Theta_{[\cdot,t]} \leq 0$, for all $i \in \{1, \ldots, |\mathcal{R}|\}$, follows directly from the structure of $\Theta_{[\cdot,t]}$ and Equation 5.59. ⋄

The correctness of U-G-RUN SCP is an immediate consequence of part (i) of Theorem 5.8 and Theorem 5.7. Furthermore, similar to the case of G-RUN implementations on totally controllable nets, an efficient implementation of U-G-RUN SCP, under a given resource ordering o and community Ψ, can be obtained by solving the LP formulation defined by Equations 5.25,5.57-5.60. This implementation of the U-G-RUN SCP is demonstrated by the following example.

Example. Consider the process-resource net depicted in Figure 5.19, that corresponds to a DIS-CON-RAS of 3 resource types, R_1, R_2 and R_3, with respective capacities 4, 3 and 4. In its current configuration, the system supports two process types, Π_1 and Π_2, with corresponding processing stage sets

$\mathcal{S}_1 = \{\Xi_{11}, \Xi_{12}, \Xi_{13}\}$ and $\mathcal{S}_2 = \{\Xi_{21}, \Xi_{22}, \Xi_{23}, \Xi_{24}\}$. The flow logic for these two processes is defined by the structure of the process sub-nets depicted in Figure 5.19. The resource allocation requests posed by the various processing stages are also indicated by the connectivity of the resource places in the net of Figure 5.19, and they are as follows: $A_{11} = (2,0,0)^T$, $A_{12} = (0,1,0)^T$, $A_{13} = (1,0,2)^T$ and $A_{21} = (1,0,1)^T$, $A_{22} = (0,2,0)^T$, $A_{23} = (0,1,0)^T$, $A_{24} = (0,0,2)^T$. In addition, the transition from stage Ξ_{22} to stage Ξ_{24} is uncontrollable; in Figure 5.19, this fact is depicted by denoting the corresponding transition with an empty rather than a filled bar.

Formulating and solving the LP defined by Equations 5.25, 5.57-5.60, for the resource ordering $o(R_1) = 2$, $o(R_2) = 1$, $o(R_3) = 3$ and community $\Psi = \{(p_{11}, p_{12}), (p_{12}, p_{13}), (p_{21}, p_{22}), (p_{22}, p_{23})\}$, leads to the following U-G-RUN implementation for the considered process-resource net:

$$\begin{pmatrix} 2 & 1 & 1 & 1 & 0 & 0 & 0 \\ 0 & 1 & 0 & 0 & 2 & 1 & 0 \\ 2 & 2 & 2 & 2 & 2 & 0 & 2 \end{pmatrix} \cdot M_S \leq \begin{pmatrix} 4 \\ 3 \\ 4 \end{pmatrix} \tag{5.62}$$

The acceptability of the LES of Equation 5.62 with respect to the uncontrollability of the net of Figure 5.19 is manifested by the fact that $A_{[\cdot, p_{22}]} \geq A_{[\cdot, p_{24}]}$. This effect is achieved by introducing a reservation of 2 units of resource R_3 at place p_{22}, which results in an additional reservation of 1 unit of R_3 for place p_{21}, in order to maintain the liveness of the controlled net.

5.2 G-RUN implementation for DIS-CON-RAS with Type-2 Uncontrollability

The Extended DIS-CON-RAS. According to the discussion of Section 5.1, in Chapter 2, Type-2 uncontrollability relates to the structuring of the sequential logic defining the various process flows supported by the system, and it is fundamentally different from the Type-1 uncontrollability, which relates to the event timings. More specifically, Type-2 uncontrollability implies the uncontrolled selection of a certain routing option by a particular process instance, due to inherent process dynamics, like an emerging need for a special process treatment or rework. From the perspective of liveness-enforcing supervision, Type-2 uncontrollability gives rise to non-live behavior that *cannot* be interpreted through the concepts of empty and deadly marked siphon, which were identified as the main cause of non-liveness in PN's modelling totally controllable RAS. In this new regime, a process instance executing a processing stage corresponding to some place $p \in P_S$ will not be able to advance to a subsequent stage as long as the (externally) selected transition $t \in p^\bullet$ is disabled, even though there might be other enabled transitions $t' \in p^\bullet$. An additional complication regarding the structural interpretation of the non-live behavior of RAS with Type-2 uncontrollability arises from the potential presence of inter-

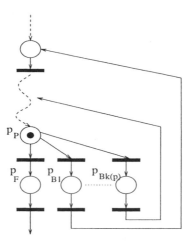

Figure 5.20.   PN-based modelling of uncontrollable process routings and reworks in the operational context of DIS-CON-RAS

nal cycles in their PN process subnets, modelling repetitive processing due to the need for rework. Yet, this section establishes that, in spite of the aforementioned modelling and analysis complications, the modelling of DIS-CON-RAS with uncontrollable process routings and potential rework requirements can be performed, for the most practical cases, through a special PN structure, which, when introduced in the original process-resource net modelling DIS-CON-RAS, enables the liveness-enforcing supervision of the resulting system through an appropriately parameterized (U-)G-RUN LES.

The key idea underlying the proposed modelling approach is the separation of the internal process dynamics determining the uncontrollable routing of a certain process instance and its potential need for rework, from the dynamics of the resource allocation concerning its advancement among the various processing stages. In the PN-based modelling framework, this separation is implemented by modelling each process stage involving uncontrolled routing for its completed process instances, through the PN subnet depicted in Figure 5.20. In the depicted subnet, tokens in place p_P correspond to process instances still in execution of the considered processing stage; tokens in place p_F correspond to process instances that have completed successfully the considered stage and request transfer to a next one; finally, tokens in places p_{B_l}, $l = 1, \ldots, k(p)$, correspond to process instances that failed the considered processing stage in a certain manner, and therefore, they request re-routing, possibly to one of the prior processing stages. Notice that any "rework" loop involves only a *single* place in P_B. Furthermore, the above discussion implies that

$$^\bullet(p_P{}^\bullet \cap {}^\bullet p_F) \cap P_R = \emptyset \tag{5.63}$$

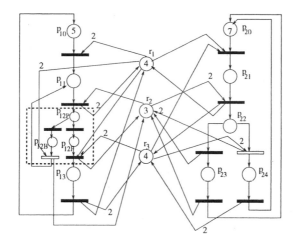

Figure 5.21.   Example: A process-resource net modelling an E-DIS-CON-RAS

since the advancement of a process instance from place p_P to p_F is a logical event, and, as such, it does not involve the allocation of any additional resources. In fact, under the aforementioned separation principle, a similar claim can be made for the transitions modelling the process advancement from place p_P to any of the places p_{B_l}, $l = 1, \dots, k(p)$; however, the proposed model allows for a non-zero resource allocation involved with these transitions, in order to accommodate the cases where a failing process instance will need some salvage and/or preparatory treatment before the actual rework stage. The resulting RAS class is characterized as *E(xtended)-DIS-CON-RAS*. The next example elucidates the structure of process-resource nets modelling E-DIS-CON-RAS, by introducing the possibility for a rework requirement in the process-resource net depicted in Figure 5.19.

Example.   Suppose that the process stage p_{12} in the process-resource net of Figure 5.19 has a non-prefect yield, and the defective outcomes must be reworked, starting from stage p_{11}. This effect is modelled by substituting the process place p_{12}, in Figure 5.19, with a subnet of the structure depicted in Figure 5.20, for $k(p) = 1$. The resource allocation vectors associated with the places in this subnet are all equal to $(0, 1, 0)^T$, i.e., the original resource allocation vector corresponding to stage p_{12}. The resulting net is depicted in Figure 5.21. ⋄

G-RUN implementation for process-resource nets modelling E-DIS-CON-RAS.   A careful study of the role of the policy community, Ψ, in the proof of Theorem 5.7, that establishes the correctness of G-RUN SCP, will reveal that the place pairs $(p, q) \in \Psi$ essentially define, for any given stage $p \in P_S$ with $p^{\bullet\bullet} \cap P_0 = \emptyset$, the set of successor stages, $q \in p^{\bullet\bullet}$, that are *guaranteed* by the policy to be accessible from p. In the case of process-resource nets modelling E-DIS-CON-RAS, tokens in places $p \in P_P$ might need to be routed to some place $q \in p^{\bullet\bullet} \cap P_B$, and therefore, these forced transitions must be guaranteed by the policy realization. This guarantee is established, in the light of the previous remark, by requiring that

$$\forall p \in P_P, \ \forall q \in p^{\bullet\bullet} \cap P_B, \ (p, q) \in \Psi \qquad (5.64)$$

In addition, the applied G-RUN realization must guarantee the progress of all processes that have completed successfully their current processing stage, i.e.,

$$\forall p \in P_S \cup P_P \cup P_F \cup P_B, \ \exists q \notin p^{\bullet\bullet} \cap P_B, \ (p, q) \in \Psi \qquad (5.65)$$

The next theorem establishes the correctness of the resulting policy.

THEOREM 5.9 *The controlled net* $\mathcal{N}^c = (P_0 \cup P_S \cup P_P \cup P_F \cup P_B \cup P_R \cup P_W, T, W, M_0)$ *corresponding to a G-RUN implementation on an E-DIS-CON-RAS for a community* Ψ *that observes the requirements of Equation 5.64 and 5.65, is live.*

The proof for Theorem 5.9 is similar to the proof of Theorem 5.7, and it can be found in (Park, 2000). Furthermore, (Park, 2000) also shows that the theory of Section 5.1, regarding the G-RUN implementation on DIS-CON-RAS with Type-1 uncontrollability, extends immediately to E-DIS-CON-RAS with uncontrollable resource allocations; the only necessary modification is the addition of Constraints 5.64 and 5.65 in the original definition of the U-G-RUN LES. The next example employs this extended theory in order to compute a U-G-RUN LES for the process-resource net of Figure 5.21, that presents, both, Type-1 and Type-2 uncontrollability.

Example.   Suppose that in the process-resource net depicted in Figure 5.21, $T_u = p_{12B}{}^{\bullet} \cup {}^{\bullet}p_{24}$. Then, implementation of U-G-RUN SCP on this net, under the resource ordering $o: o(R_1) = 2, \ o(R_2) = 1, \ o(R_3) = 3$, and community $\Psi = \{(p_{11}, p_{12}), (p_{12P}, p_{12F}), (p_{12P}, p_{12B}), (p_{12F}, p_{13}), (p_{12B}, p_{11}), (p_{21}, p_{22}), (p_{22}, p_{23})\}$, leads to the following set of constraints:

$$\begin{pmatrix} 2 & 2 & 1 & 2 & 1 & 1 & 0 & 0 & 0 \\ 0 & 1 & 1 & 1 & 0 & 0 & 2 & 1 & 0 \\ 2 & 2 & 2 & 2 & 2 & 2 & 2 & 0 & 2 \end{pmatrix} \cdot M_S \leq \begin{pmatrix} 4 \\ 3 \\ 4 \end{pmatrix} \qquad (5.66)$$

In Equation 5.66, the rows of the policy-defining matrix, A, correspond to the resource sequence $< R_1, R_2, R_3 >$, and its columns correspond to the place sequence $< p_{11}, p_{12P}, p_{12F}, p_{12B}, p_{13}, p_{21}, p_{22}, p_{23}, p_{24} >$. The juxtaposition of the supervisor of Equation 5.66 with that of Equation 5.62 indicates that the primary effect of the introduced rework loop is to increase the effective requirement of stage p_{12} with respect to resource R_1 by one unit. This increase hedges against the possibility of re-executing stage p_{11}, and the extra resource unit is released, once a positive process outcome is established.

5.3 Liveness-Enforcing Supervisors for CPX-RAS with uncontrollable behavior

The results of Sections 5.1 and 5.2 can be extended towards the development of liveness-enforcing supervisors for the broader class of CPX-RAS with Type-1 and/or Type-2 uncontrollability, through a problem reduction technique similar to that employed in Section 3.2. A first step in this approach is to develop an acceptable supervisor Δ_j for each process type Π_j that can support the nonblocking execution of any single process instance of that type, in isolation. These supervisors can be obtained by applying the algorithm of Figure 2.9 on the STD's modelling the reachability spaces of the corresponding resource-augmented process subnets, where the latter are appropriately embellished in order to express the involved uncontrollability. Under the assumption that the process behavior enabled by each supervisor Δ_j is non-vacuous, the control logic enforced by each supervisor Δ_j can be readily encoded to a resource-augmented process net of the (E-)DIS-CON-RAS type, and the availability of these nets subsequently enables the synthesis of a G-RUN-based liveness-enforcing supervisor for the entire RAS, through the theory developed in Sections 5.1 and 5.2. We leave to the reader the details of the implementation of this approach. We notice, however, that, as in the case of Section 3.2, the resulting supervisor will be of non-polynomial complexity with respect to the underlying RAS size. A thinning procedure can be applied to each process-controlling supervisor, Δ_j, in this case, too. However, the implementational details of this procedure are further complicated by the requirement that the returned supervisors must be acceptable with respect to the uncontrollability inherent in the underlying resource-augmented process subnets.

6. Historical and bibliographical notes

The PN-based modelling framework has always been one of the main frameworks employed for the modelling, analysis and control of the RAS classes considered in this work. For instance, as it was mentioned in Chapter 4, the work of (Banaszak and Krogh, 1990), that constitutes one of the very first works to deal with the problem of developing PK-SCP's for the class of LIN-SU-RAS, was developed using a PN-based model. As it is manifested by the results presented in this chapter, PN's allow for the explicit and rigorous characterization of the underlying system structure and its impact on the generated behavior, and in this way, they complement the analytical and computational capability provided by the FSA-based modelling, that focuses more on the behavioral patterns generated by the studied system.

The first results to connect the lack of liveness and reversibility of process-resource nets to the structural concept of empty siphon can be traced in the seminal work of Ezpeleta and his colleagues (Ezpeleta et al., 1995). More specifically, (Ezpeleta et al., 1995) established that in the PN class modelling DIS-SU-RAS, non-liveness can be interpreted through the development of *empty* siphons during the system operation. Subsequently, the work of (Chu and Xie, 1997) established that the presence of reachable empty siphons is also the cause for non-liveness in *Augmented Marked Graph (AMG)*, a class of PT-ordinary PN's corresponding to COR-CON-RAS where resources are allocated and de-allocated in bundles containing up to a single unit from each resource. Another important contribution of the work presented in (Chu and Xie, 1997) was the development of a sufficiency test for the non-existence of reachable empty siphons, that takes the convenient form of a mathematical programming (MP) formulation polynomially sized, in terms of variables and constraints, with respect to the underlying PN model. The role of empty siphons for some additional RAS classes modelled by PT-ordinary PN's was further investigated in (Xie and Jeng, 1999). At the same time, the work of (Park and Reveliotis, 2000) established that the behavior of a LIN-SU-RAS under the control of *algebraic* LES can be modelled as an AMG, and therefore, the aforementioned results of (Chu and Xie, 1997) provide a structural test for assessing the LES correctness.

A complete siphon-based characterization of the (non-)liveness of process-resource nets modelling DIS-CON-RAS can be found in (Park and Reveliotis, 2001b), while some earlier attempts can be found in (Barkaoui and Ben Abdallah, 1996; Barkaoui et al., 1997; Tricas et al., 1998; Tricas et al., 1999). This characterization was subsequently generalized for CPX-CON-RAS modelled by process-resource nets with acyclic, quasi-live and strongly reversible process subnets in (Reveliotis, 2003a), that constitutes the basis of the material provided in Sections 1 and 2.1. In addition, the work of (Park and Reveliotis, 2001b) developed the computational tools presented in 2.2, for verifying the liveness of process-resource nets with acyclic, quasi-live and strongly reversible process subnets, and demonstrated that, in the case of algebraic SCP's, these tools can also function as policy correctness verification tools. Another line of work that was developed simultaneously with the work reported in (Reveliotis, 2003a), and presents strong similarity to it, even though it remains in the domain of process-resource nets modelled by PT-ordinary PN's, is that presented in (Jeng et al., 2002). An additional important contribution of (Jeng et al., 2002) is the structural characterization of the property of strong reversibility, that was introduced by Assumption 3, and the development of a computational test for its verification. A more extensive treatment of the topic of defining and designing well-formed process subnets is provided by *workflow* theory; c.f., (Van der Aalst, 1996; Van der Aalst, 1997; Van der Aalst and Van Hee, 2002; Van Hee et al., 2003) for an introduction to it. Some additional results concerning the role of siphons in the liveness and liveness-enforcing supervision of any general PN can be found in (Iordache and Antsaklis, 2003). Finally, the work of (Fanti et al., 2000) investigated the correspondence between the empty siphon interpreting a LIN-SU-RAS deadlock in the PN-based representation, and the cyclical resource dependencies corresponding to the same deadlock, under a graph-theoretic representation of the RAS state similar to those employed in Chapters 3 and 4.

The synthesis of LES for CPX-RAS through reduction to DIS-CON-RAS was first proposed in (Chew et al., 2003). The extension of the RUN SCP so that it applies to DIS-CON-RAS, was pursued in (Reveliotis et al., 1997; Park et al., 2001; Park and Reveliotis, 2001b; Park and Reveliotis, 2001a; Park, 2000); the material of Section 4.2 is based on (Park and Reveliotis, 2001a; Park, 2000). The material of Sections 5.1 and 5.2, regarding the G-RUN implementation on DIS-CON-RAS with Type-1 and Type-2 uncontrollability is based on the results of (Park, 2000; Park and Reveliotis, 2002a). The material presented in Section 4.1, regarding the application of Banker's algorithm in the DIS-CON-RAS operational context, is a straightforward extension of the relevant results on Banker's algorithm presented in Chapter 4. Another work considering the implementation of Banker's algorithm in an extended DIS-CON-RAS context, where the various processes can possess controllably cycling routes, is that presented in (Ezpeleta et al., 2002).

We conclude the discussion of this section by noticing that the current literature provides some additional approaches to the problem of liveness-enforcing supervision arising in the operational context of the considered RAS classes. Generally speaking, all these approaches try to identify and disable the occurrence of all the unsafe states that can take place in the system, based on complete or partial enumeration of these states, and therefore, they result in LES's that present super-polynomial complexity with respect to the size of the underlying RAS. In addition, they tend to ignore the problems that could arise from this increased complexity, and as a result, they fail to provide any additional mechanisms for their effective management. Hence, while such approaches can possibly lead to highly permissive supervisors, their applicability is limited by the fact that they are non-scalable. Based on this observation, we opted to not include them in the problem treatment presented in this work. The interested reader can trace some of these approaches in, e.g., (Ezpeleta et al.,

1995; Barkaoui and Ben Abdallah, 1995; Yalcin and Boucher, 2000; Ghaffari et al., 2003; Zhou and Fanti, 2004).

Chapter 6

PERFORMANCE-ORIENTED MODELLING AND CONTROL OF LOGICALLY CONTROLLED RAS

In this chapter we shift attention to the second major component of the control framework depicted in Figure 1.7, addressing the performance-oriented control problem for RAS that are supervised by one of the logical control policies developed in the previous chapters. As it was pointed out in the discussion of the control framework of Figure 1.7, the logical control function essentially constitutes a *filter* that determines, for any given RAS state, the subset of admissible actions among the broader set of feasible actions at that state. It is the performance-oriented controller that determines the particular admissible action to be commanded upon the system. In the RAS operational context, typical actions concern the advancement of a process instance that has completed its current processing stage to one of its successor stages together with the allocation of the necessary resources. Hence, this performance-oriented control problem belongs to the class of problems known as *scheduling* problems (Pinedo, 2002).

The detailed formulation of a scheduling problem depends on the adopted operational assumptions for the underlying resource allocation system and the performance objective(s) to be observed. Regarding the latter, typical considerations include the maximization of the system throughput, the minimization of some inventory cost function, and the minimization of the process tardiness with respect to some target deadlines, known as the process due dates. It is also well-established that most of the resulting formulations with practical applicability are NP-Hard problems (Pinedo, 2002; Garey and Johnson, 1979). In the light of this negative result, scheduling decisions have been addressed in practice through a class of heuristics known as *dispatching rules* (Pinedo, 2002; Nahmias, 1997; Van der Aalst and Van Hee, 2002). Generally speaking, dispatching rules seek to arbitrate the resource allocation contest among the various competing processes on the basis of simple process attributes like the number of their remaining processing stages, their remaining time to completion, the proximity of their due date, etc. Different rules are associated with different performance objectives, and their suitability for these objectives is founded on experimental evidence, practical experience, and, in certain cases, their proven optimality for some simpler problem versions (Pinedo, 2002; Kumar, 1994a; Kumar, 1994b). An extensive classification and evaluation of the most frequently used dispatching rules can be found in (Panwalkar and Iskander, 1977; Blackstone et al., 1982).

It should be obvious to the reader that the dispatching rule-based approach to scheduling is also immediately applicable to the RAS performance-oriented control problem addressed in this chapter: at every decision cycle of the control framework depicted in Figure 1.7, this approach will prioritize the set of admissible actions based on the criteria established by the applied dispatching rule, and it will select the action with the highest priority. Hence, the framework of Figure 1.7 and the associated logical control policies are totally compatible with the mainstream scheduling practice. However, in this chapter we seek to provide a modelling framework that will enable the systematic characterization of the optimal policy for the RAS scheduling problem formulated under the operational assumptions of Chapter 1, while observing some of the most typically addressed performance objectives. The availability of such a characterization will provide a more profound understanding of the nature and structure of the optimal scheduling policy, and it is expected to eventually enable the development of effective and computationally efficient approximations to it. It will also facilitate the systematic characterization of the impact on the system performance of any decisions pertaining to the underlying RAS structure and the applied logical control policy, allowing, thus, the further rationalization of these decisions. Each of the above items constitutes the subject of a corresponding section. In general, the problem of performance-oriented control for the considered RAS class has not been studied as extensively as its counterpart problem pertaining to the logical control of these environments. Hence, the intention of this chapter is to highlight the state of art with respect to this topic, but also, to outline and motivate opportunities for further research activity in this area.

1.   Performance-oriented modelling and control of logically controlled RAS

In order to analyze and eventually control the performance of logically controlled RAS with respect to the aforementioned criteria of maximizing the system throughput and / or minimizing the process inventory cost and tardiness, we must augment the previously developed models, that provided qualitative, logical characterizations of the RAS behavior, with additional elements able to capture its timed dynamics. The resulting modelling frameworks belong to the realm of *Timed Automata* and *Stochastic Processes* (Cassandras and Lafortune, 1999). Following a pattern similar to that adopted in the qualitative modelling of the RAS dynamics, the first part of this section briefly introduces the afore-mentioned modelling frameworks and some key results that are necessary for the formulation and analysis of the RAS performance-oriented control problem; the detailed modelling of the RAS timed dynamics and the ensuing formulation of the RAS performance-oriented control problem are presented in the second part of the section.

1.1 Timed Models of Discrete Event Systems

In this section, we present a set of models that can be employed for the modelling and analysis of the timed dynamics generated by the broader class of Discrete Event Systems (DES). The presented models essentially constitute part of a taxonomy, with the most encompassing members of the taxonomy presenting higher modelling power but also reduced analytical tractability. Our presentation proceeds from the most general models of this class to the most constrained, seeking to systematically reveal their inter-relationships, and the particular set of assumptions that defines each model as a restriction of its parent model. The explicit identification of these assumptions is crucial for assessing the model applicability, understanding its limitations, and, possibly, adjusting the original model in order to serve better the needs of the pursued analysis. Our development of this material is primarily based on the exposition of (Cassandras and Lafortune, 1999).

Stochastic Processes

A *stochastic* or *random process* $X(\omega, t)$ is a collection of random variables indexed by a variable t that ranges over some given set $T \subseteq \Re$. These random variables are defined over a common *probability space*, (Ω, E, P), where Ω denotes the set of possible outcomes, alternatively known as the *sample space*, $E \subseteq 2^{\Omega}$ denotes the *event space*, and P is the *probability measure* defined over the event set E. In the particular case that $X(t)$, $t \in T$, takes values over a finite or countable set, the stochastic process is characterized as a *discrete-state* process, or a *chain*; otherwise, it is a *continuous-state* process. In most applications of stochastic processes, the variable t is modelling the passage of time. Hence, if the set T is finite or countable, the process is said to be *discrete-time*; otherwise, it is *continuous-time*.

A stochastic process $\{X(t)\}$ is fully characterized by the *joint cumulative distribution function (cdf)*:

$$F_X(x_0, \ldots, x_n; t_0, \ldots, t_n) = P(X(t_0) \leq x_0, \ldots, X(t_n) \leq t_n) \qquad (6.1)$$

for all possible tuples of $(x_0, \ldots, x_n)$ and $(t_0, \ldots, t_n)$. Of particular interest in the subsequent developments is a subset of stochastic processes $\{X(t)\}$ with a joint cdf F_X that is invariant to translation in time, i.e.,

$$F_X(x_0, \ldots, x_n; t_0 + \tau, \ldots, t_n + \tau) = F_X(x_0, \ldots, x_n; t_0, \ldots, t_n), \quad \forall \tau \in \Re$$
(6.2)

These stochastic processes are called *stationary*, since their stochastic behavior does not depend on the particular interval of observation.

Markov and Semi-Markov Processes

Two other important stochastic process subclasses that are obtained by constraining the structure of the joint cdf of Equation 6.1 are those of *Markov* and *Semi-Markov* processes.

Markov processes.   A stochastic process is said to be a *Markov* process *iff*

$$P(X(t_{k+1}) \leq x_{k+1} \mid X(t_k) = x_k, X(t_{k-1}) = x_{k-1}, \ldots, X(t_0) = x_0) = $$
$$P(X(t_{k+1}) \leq x_{k+1} \mid X(t_k) = x_k) \quad (6.3)$$

for any $t_0 \leq t_1 \leq \ldots \leq t_k \leq t_{k+1}$. Discrete-state Markov processes are known as *Markov chains*, and, in their case, Equation 6.3 takes the simpler form:

$$P(X(t_{k+1}) = x_{k+1} \mid X(t_k) = x_k, X(t_{k-1}) = x_{k-1}, \ldots, X(t_0) = x_0) = $$
$$P(X(t_{k+1}) = x_{k+1} \mid X(t_k) = x_k) \quad (6.4)$$

In words, Equations 6.3 and 6.4 imply that the future evolution of the process depends only on its current state, which adequately summarizes all its past history. This effect has come to be known in the relevant literature as the *"memoryless" property* of Markov processes, and it has two important implications for the dynamics of the underlying system, stated in the following two assumptions:

ASSUMPTION 6 *All past state information is irrelevant for the future evolution of the process.*

ASSUMPTION 7 *The length of the process sojourn in its current state is irrelevant for the future evolution of the process.*

It can be shown (Cassandras and Lafortune, 1999) that Assumption 7 further implies that

ASSUMPTION 8 *The process sojourn time at any visited state is* exponentially *distributed.*

Clearly, Assumptions 7 and 8 restrict drastically the set of physical processes that can be modelled as Markov processes: In most contemporary technological applications, processing times are more deterministically distributed around a certain mean value, and therefore, knowledge of the elapsed time since the initiation of a certain process provides some information regarding the remaining time to its completion. However, as it will be shown in a subsequent section, it is possible to approximate most processing time distributions of practical interest, to any degree of accuracy, by means of a continuous-time stochastic process that retains all the important elements of the Markovian behavior described above. This is really fortunate, since Markov chains is the main subclass of stochastic processes that is analytically tractable.

More specifically, the dynamics of a Continuous-Time Markov chain (CTMC) with a state space $\mathcal{X}$, can be completely defined by (i) its *infinitesimal generator* $Q(t)$, and (ii) a probability distribution $\pi_0(x)$, $x \in \mathcal{X}$, characterizing the process initial state x_0. The process infinitesimal generator $Q(t)$ is a $|\mathcal{X}| \times |\mathcal{X}|$ matrix with its non-diagonal elements $q_{ij}(t)$ denoting the rate with which the process transitions from state x_i to state x_j[1] at time t; diagonal elements $q_{ii}(t)$ are set equal to $-\sum_{j \neq i} q_{ij}(t)$. Of particular interest in the subsequent developments is the case of $Q(t) = Q$, $\forall t$, corresponding to stationary or *(time-)homogeneous* CTMC's. The probability distribution $\pi_t(x)$, $x \in \mathcal{X}$, characterizing the state occupancy of a homogeneous CTMC at time t, is given by:

$$\pi_t^T = \pi_0^T \cdot e^{Qt} \tag{6.5}$$

where $e^{Qt} \equiv \sum_{i=0}^{\infty} Q^i t^i / i!$. Of special interest in the mathematical theory of Markov chains, but also, in our subsequent developments, is the behavior of the vector π_t as $t \to \infty$. This behavior is conditioned upon the presence or absence of some additional structure in the chain state space $\mathcal{X}$, which is discussed next.

Irreducible, homogeneous, positive recurrent CTMC's and their steady-state distribution. The availability of the process infinitesimal generator Q, defined above, allows also the determination of (i) the exponential distribution that governs the sojourn times that are experienced by the process at any state $x_i \in \mathcal{X}$, and (ii) the probability distributions that govern the transition of the process out of any state x_i. More specifically, the rate of the exponential distribution characterizing the process sojourn times experienced at state x_i, is computed by

$$\Lambda_i = \sum_{j \neq i} q_{ij} \tag{6.6}$$

while the *expected* process sojourn time at state x_i is equal to

$$\tau_i = 1/\Lambda_i = 1/(\sum_{j \neq i} q_{ij}) \tag{6.7}$$

---

[1] Unless indicated otherwise, in the following, it is assumed that the indexing of the infinitesimal generator Q corresponds to the indexing of the process states.

The probabilities p_{ij} for the process transition from state x_i to state x_j, given that it is in state x_i, – known as the process *branching probabilities*, in the relevant literature – are computed by

$$p_{ij} = q_{ij}/(\sum_{j \neq i} q_{ij}) \tag{6.8}$$

The branching probabilities p_{ij} induce a Discrete-Time Markov chain (DTMC) on the state space $\mathcal{X}$ of the original CTMC, which can be employed towards the further characterization of the process behavior in various parts of its state space. Hence, a state x_j of $\mathcal{X}$ is said to be *reachable* from some other state x_i, *iff* the aforementioned DTMC can find itself in state x_j while starting its operation from state x_i. A subset S of $\mathcal{X}$ is said to be *closed*, *iff* $p_{ij} = 0$ for any $x_i \in S$, $x_j \notin S$; in particular, a state x_i that is forming a single-element closed set, is said to be *absorbing*. A closed set of states S is said to be *irreducible*, *iff* any two states of it are reachable from each other. A Markov chain is said to be *irreducible*, *iff* its entire state space $\mathcal{X}$ is irreducible.

It is clear from the above definitions, that a closed state set S is a *"trapping"* set for the process behavior; once the process finds itself in S, it will stay in it forever. In the particular case that S is also irreducible, the underlying strong connectivity of S implies that every state in it will be revisited an infinite number of times, as $t \to \infty$; hence, states belonging to irreducible subsets of the process state space, $\mathcal{X}$, are said to be *recurrent*. It can also be shown that, in the particular case that the considered irreducible subset S is finite, the expected time between any two consecutive revisits to any given state of S will be finite, and therefore, the states of S are said to be *positive recurrent*. States of $\mathcal{X}$ that are not recurrent, are said to be *transient*; for these states, the probability of returning to them is strictly less than one.

Now we are ready to state a key result regarding the "long-term" behavior of the CTMC state occupancy probability vector, π_t, for $t \to \infty$:

THEOREM 6.1 *In an irreducible, homogeneous CTMC with positive recurrent states, the state occupancy probability vector, π_t, converges componentwise, as $t \to \infty$, to a vector π, that is known as the* steady-state *or* equilibrium *probability vector. Furthermore, the limit vector π can be computed as the* unique *solution to the following system of linear equations:*

$$\pi^T \cdot Q = 0 \tag{6.9}$$

$$\sum_{x \in \mathcal{X}} \pi(x) = 1 \tag{6.10}$$

Since all the states $x \in \mathcal{X}$ of an irreducible, homogeneous CTMC with $|\mathcal{X}| < \infty$, are positive recurrent, we have also the following interesting corollary of Theorem 6.1:

COROLLARY 6 *Irreducible, homogeneous CTMC's, with a* finite *state space* $\mathcal{X}$, *possess a steady-state distribution* π, *that is provided by Equations 6.9 and 6.10.*

This corollary will be very useful in the subsequent developments.

Semi-Markov processes. A stochastic process is said to be *semi-Markov*, if it satisfies only Assumption 6. More specifically, a semi-Markov process is a continuous-time stochastic process $\{X(t),\ t \in \mathcal{T}\}$, such that

1 the embedded discrete-time stochastic process $\{X_n,\ n \in \mathcal{N}\}$, defined by the observation of the original process $\{X(t),\ t \in \mathcal{T}\}$ upon each transition epoch, is a discrete-time Markov chain;

2 the process sojourn time in each state is a continuous random variable or a deterministic duration, and these sojourn times are mutually independent (Viswanadham and Narahari, 1992).

We notice, for completeness, that the long-term behavior of semi-Markov processes can be studied through the steady-state analysis of the aforementioned embedded DTMC $\{X_n,\ n \in \mathcal{N}\}$; since, however, this set of results is not immediately relevant to the subsequent developments, we refer to (Viswanadham and Narahari, 1992) for the details.

Stochastic Timed Automata

Next we link the theory of stochastic processes outlined above, to the modelling framework of finite state automata, that was used in the previous chapters as a primary modelling framework for the RAS logical analysis. The resulting modelling framework is that of the *stochastic timed automaton*, and it combines the structural and behavioral elements of FSA's with the timed dynamics of stochastic processes. A formal definition of the stochastic timed automaton, that serves the needs of this book, is as follows:

DEFINITION 26 *(Cassandras and Lafortune, 1999) A stochastic timed automaton is a six-tuple*

$$(\mathcal{E}, \mathcal{X}, \Gamma, p, p_0, \mathcal{C})$$

where

- $\mathcal{E}$ *is a finite* event set*;*

- $\mathcal{X}$ *is a finite[2] state space;*

- $\Gamma(x)$, $x \in \mathcal{X}$, *is the set of* feasible *or* enabled *events in state x;*

- $p(x'; x, e)$, $x, x' \in \mathcal{X}, e \in \mathcal{E}$, *is the* state transition probability *function,* satisfying $p(x'; x, e) = 0$, for all $e \notin \Gamma(x)$;

- $p_0(x)$, $x \in \mathcal{X}$, *defines the probability* $P(X_0 = x)$ *for the initial state* X_0;

- $\mathcal{C} = \{\mathcal{C}_e, e \in \mathcal{E}\}$ *is a* stochastic clock structure, *associating with every event* $e \in \mathcal{E}$ *a distribution* $\mathcal{C}_e$ *that characterizes the random sequences of lifetimes,* $\{V_{e,k}, k = 1, 2, \ldots, \}$ $V_{e,k} \in \Re^+$, *corresponding to event e.*

The aforementioned automaton generates a stochastic sequence $\{X_n\}, n = 0, 1, \ldots$, through the following transition mechanism: Upon the entrance of the automaton in some state x, the *residual lifetimes*, Y_e, of the events $e \in \Gamma(x)$ are compared, and the event e^* with the smallest residual lifetime Y_{e^*} is the one to be executed next, after the passage of Y_{e^*} time units. The execution of event e^* results in the transition of the automaton from state x to some state x', with probability $p(x'; x, e^*)$. At this point, the residual lifetimes, Y_e', for events $e \in \Gamma(x')$ are determined, and the entire process repeats itself. Regarding the updating of the residual lifetimes, we notice that the residual lifetimes, Y_e', for events $e \in \Gamma(x) \cap \Gamma(x')$, $e \neq e^*$, are determined by $Y_e' = Y_e - Y_{e^*}$, i.e., by reducing appropriately the earlier residual lifetimes in order to account for the time spent by the automaton in state x. For any other event $e \in \Gamma(x')$, the residual lifetime Y_e' is obtained as the next value in the sequence $\{V_{e,k}\}$; this sequence is generated through sampling of the distribution $\mathcal{C}_e$. Finally, as mentioned in Definition 26, the automaton initial state X_0 is selected from $\mathcal{X}$ through sampling with probability distribution p_0, and the residual lifetimes for the events $e \in \Gamma(X_0)$ are set to $Y_e = V_{e,1}$.

The stochastic sequence $\{X_n\}$ can be naturally extended to a continuous-time stochastic process $\{X(t), t \in \Re_0^+\}$ by setting $X(t) = X_n$, $\forall t \in [t_n, t_{n+1})$, $n = 0, 1, \ldots$, where t_n denotes the occurrence time of the n-th transition. Process $\{X(t)\}$ is known as the *Generalized semi-Markov process (GSMP)* generated by the stochastic timed automaton $(\mathcal{E}, \mathcal{X}, \Gamma, p, p_0, \mathcal{C})$. While GSMP's characterize completely the timed dynamics underlying the operation of their generating stochastic automata, in the general case, they are too complex to be analytically tractable. Therefore, they are primarily studied through simulation of the automaton operation according to the logic outlined above.[3] However, in the

---

[2]Notice that, in the more general definition of the stochastic timed automaton presented in (Cassandras and Lafortune, 1999), sets $\mathcal{E}$ and $\mathcal{X}$ are allowed to be countable.

[3]In fact, the modelling abstraction of the stochastic automaton can be considered as a formalization of the structure of the typical discrete-event simulator.

particular case that the lifetime sequences $\{V_{e,k}\}$, for each event $e \in \mathcal{E}$, consist of mutually independent random variables, identically distributed according to an exponential distribution with rate λ_e, the aforementioned GSMP structure reduces to a CTMC with the corresponding infinitesimal generator, Q, being defined as follows:

$$\forall i,j, \text{ with } i \neq j, \quad q_{ij} = \sum_{e \in \Gamma(x_i)} p(x_j; x_i, e)\lambda_e \qquad (6.11)$$

Once the infinitesimal generator has been defined, the process dynamics can be analyzed according to the earlier presented CTMC theory.

Markov Decision Processes

The previously described models of stochastic processes belong to the class of *"descriptive"* models, since their main role is to characterize the timed-dynamics, and thus, enable the evaluation of the performance of any given system configuration. Since our ultimate objective is the control of the system behavior in order to meet some performance criterion, we need an additional class of models that will aid us in the determination of the control function; models with this capability are characterized as *"prescriptive"*. In the context of stochastic processes and automata, the most well-established prescriptive model is that of a *Markov Decision Process (MDP)* (Puterman, 1994). Generally speaking, MDP models seek to regulate the occurrence rate of the various events / state transitions during the system operation so that some pre-specified performance objective is optimized. Next we present a particular variation of the MDP model and some results associated with it that will enable the modelling of the RAS performance-control problem undertaken in the following section.

Continuous-Time MDP's. The MDP model that is considered in the rest of this chapter belongs to the class of MDP models known as *Continuous-Time MPD's (CTMDP)*. A formal definition of CTMDP models is as follows:

DEFINITION 27 *A continuous-time Markov Decision Process (CTMDP) is defined as a six-tuple $(\mathcal{X}, \mathcal{U}, \Gamma, \Lambda, p, g)$, where:*

- $\mathcal{X}$ *is the finite[4] state space of the underlying process.*

- $\mathcal{U}$ *is a finite set of control actions.*

- $\Gamma : \mathcal{X} \to 2^{\mathcal{U}}$ *is a function with $\Gamma(x) \neq \emptyset$ specifying the set of control actions available at state x.*

---

[4]In a more general definition of CTMDP's, $\mathcal{X}$ can also be a countable set.

- $\Lambda : \mathcal{X} \times \mathcal{U} \to \Re^+$ *is a function that returns, for any given state* x *and control action* $u \in \Gamma(x)$, *the* rate $\Lambda(x, u)$ *for the exponential distribution characterizing the process sojourn time at state* x *under the selection of control action* u; Λ *is undefined for all other state-action pairs.*

- p *is a set of probability distributions, one for each pair* $(x \in \mathcal{X}, u \in \Gamma(x))$, *with each distribution defined over the set* $\mathcal{X}$, *and with* $p(x'; x, u)$ *being the* process branching probability *to state* x', *upon executing action* u *in state* x.

- $g : \mathcal{X} \times \mathcal{U} \to \Re$ *is a function defined only for state-action pairs* (x, u) *with* $u \in \Gamma(x)$, *and with* $g(x, u)$ *being the* rate of accumulation *of some* immediate gain *or* reward *obtained by executing action* u *in state* x.

The operation of the above MDP model evolves according to the following basic cycle: Upon the entrance of the underlying stochastic process in some state $x \in \mathcal{X}$, a control action $u \in \Gamma(x)$ is selected and executed according to some decision rule d that is externally specified. Subsequently, the process remains in state x for a certain amount of time, ζ, that is determined through sampling from an exponential distribution with parameter rate $\Lambda(x, u)$, and, at the end of this interval, it transitions to another state x', according to the branching probability distribution $p(x'; x, u)$. During the interval ζ, the process also accumulates a gain with rate $g(x, u)$. Once in state x', the process repeats the cycle described above.

CTMDP policies. The key objective of an MDP problem is to determine the sequence of the decision rules d to be applied at each decision epoch so that some function of the accrued process gain is optimized. Of particular interest in the MDP theory and the subsequent developments, is the case that the rule d to be applied at each decision epoch is invariant with respect to the actual decision time or the order of the decision epoch, and it depends only on the current state x of the process; the policy Ψ resulting by such a decision rule structure is characterized as *stationary*. Stationary policies can be described, in general, by a set of *action selection probability distributions*, $\psi^\Psi(\cdot; x)$, $x \in \mathcal{X}$, defined over the corresponding control action sets $\Gamma(x)$. In the particular case that the support of each distribution $\psi^\Psi(\cdot; x)$, $x \in \mathcal{X}$, is a singleton, policy Ψ is said to be *deterministic*; otherwise, it is said to be *randomized*. In the following discussion, we shall assume that the considered policies Ψ are *stationary randomized* policies, unless otherwise specified.

CTMDP objectives. In order to completely define a CTMDP problem, we must also quantify the objective function that is to be optimized by the specification of policy Ψ. As mentioned above, this objective function is defined in terms of the gain rates $g(x, u)$ that are experienced by the underlying process. More specifically, this function typically constitutes a weighted integral of the experienced gain rates $g(x, u)$, with the various weights specified so that the resulting optimization problem is well-posed. Some objective functions typically considered in the literature, are the following:

Total expected gain over a finite horizon:

$$V^\Psi(x_0) = E_{x_0}^\Psi \left[\int_0^T g(x(t), u(t)) dt \right] \tag{6.12}$$

Total expected discounted gain over an infinite horizon:

$$V^\Psi(x_0) = E_{x_0}^\Psi \left[\int_0^\infty e^{-\beta t} g(x(t), u(t)) dt \right] \tag{6.13}$$

Expected average gain (over an infinite horizon):

$$V^\Psi(x_0) = \lim_{T \to \infty} \frac{1}{T} E_{x_0}^\Psi \left[\int_0^T g(x(t), u(t)) dt \right] \tag{6.14}$$

In Equations 6.12-6.14, the expectation $E_{x_0}^\Psi[\cdot]$ is taken over all process realizations under policy Ψ, while the process is initialized at state x_0. The MDP problem resulting from the adoption of any of these three objective functions, is the identification of a policy Ψ^* that maximizes the value of the corresponding objective function, across all states $x \in \mathcal{X}$; formally,

$$\Psi^* = \arg\sup_\Psi \left\{ V^\Psi(x), \ \forall x \in \mathcal{X} \right\} \tag{6.15}$$

The reader can convince herself that fixing the applied action selection logic to some particular stationary randomized policy Ψ, reduces the MDP dynamics to a CTMC, with the corresponding infinitesimal generator, Q^Ψ, being defined as follows:

$$\forall i, j, \text{ with } i \neq j, \ q_{ij}^\Psi = \sum_{u \in \Gamma(x_i)} \psi^\Psi(u; x_i) \Lambda(x_i, u) p(x_j; x_i, u) \tag{6.16}$$

Hence, in principle, any given stationary randomized policy Ψ can be evaluated with respect to the objective functions of Equations 6.12-6.14 by analyzing the dynamics of the corresponding CTMC.

Communicating expected average gain CTMDP problems.   Next, we focus on a class of expected average gain CTMDP problems, with the additional property that, for any given pair of states $x, x' \in \mathcal{X}$, there exists a *stationary deterministic* policy Ψ under which x' is accessible from x; CTMDP's possessing this property are characterized as *communicating*. It can be shown that for expected average gain problems defined on communicating CTMDP's, there will always exist an optimal policy Ψ^* that is *unichain*, i.e., the CTMC induced by the policy Ψ^* consists of a *single* recurrent class of states plus a possibly empty set of transient states (Puterman, 1994). Hence, under policy Ψ^*, the underlying stochastic process will eventually be confined in an irreducible subset of $\mathcal{X}$, to be denoted by $\mathcal{X}_r^{\Psi^*}$. But then, Corollary 6 implies that the long-term dynamics of this process will be characterized by a steady-state distribution, π^{Ψ^*}, obtained by solving the system of Equations 6.9 and 6.10, for the process restriction on subspace $\mathcal{X}_r^{\Psi^*}$. Let also $\mathcal{X}_t^{\Psi^*}$ denote the set of transient states under policy Ψ^*, i.e., $\mathcal{X}_t^{\Psi^*} = \mathcal{X} \backslash \mathcal{X}_r^{\Psi^*}$; π^{Ψ^*} can be extended over the entire state space $\mathcal{X}$ by setting $\pi^{\Psi^*}(x) = 0$ for all $x \in \mathcal{X}_t^{\Psi^*}$. The availability of π^{Ψ^*} subsequently allows the representation of $V^{\Psi^*}(x_0)$, for the considered problem, through the following formula:

$$V^{\Psi^*}(x_0) = \sum_{x \in \mathcal{X}} \sum_{u \in \Gamma(x)} g(x, u) \pi^{\Psi^*}(x) \psi^{\Psi^*}(u; x) \qquad (6.17)$$

Notice that an important implication of Equation 6.17 is that, for the considered class of expected average gain CTMDP problems, $V^{\Psi^*}(x_0)$ does not depend on the initial state x_0; hence, in the following, the optimal expected average gain of the considered CTMDP problems will be denoted by V^*. The characterization of V^* through Equation 6.17 also implies that V^* and an optimal policy Ψ^* can be obtained by solving the following mathematical programming formulation, in variables $\psi^{\Psi}(u; x)$ and $\pi^{\Psi}(x)$, $x \in \mathcal{X}$, $u \in \Gamma(x)$:

$$V^* = \max_{\psi^{\Psi}(u;x), \pi^{\Psi}(x)} \sum_{x \in \mathcal{X}} \sum_{u \in \Gamma(x)} g(x, u) \pi^{\Psi}(x) \psi^{\Psi}(u; x) \qquad (6.18)$$

s.t.

$$(\pi^{\Psi})^T \cdot Q(\psi^{\Psi}) = 0 \qquad (6.19)$$

$$\sum_{x \in \mathcal{X}} \pi^{\Psi}(x) = 1 \qquad (6.20)$$

$$\sum_{u \in \Gamma(x)} \psi^{\Psi}(u; x) = 1, \quad \forall x \in \mathcal{X} \qquad (6.21)$$

$$\pi^{\Psi}(x) \geq 0; \ \psi^{\Psi}(u;x) \geq 0, \ \forall x \in \mathcal{X}, \ \forall u \in \Gamma(x) \qquad (6.22)$$

The elements of the infinitesimal generator $Q(\psi^{\Psi})$, that appears in Equation 6.19, are defined by Equation 6.16, and they constitute functions of $\psi^{\Psi}(u;x)$. Hence, the structure of Equation 6.19, and also the structure of the objective function, render the formulation of Equations 6.18-6.22 non-linear. Fortunately it can be linearized, by introducing the dummy variables

$$y^{\Psi}(x,u) = \pi^{\Psi}(x)\psi^{\Psi}(u;x)\Lambda(x,u), \ \forall x \in \mathcal{X}, \ \forall u \in \Gamma(x) \qquad (6.23)$$

Then, by also setting

$$\tau(x,u) = 1/\Lambda(x,u), \ \forall x \in \mathcal{X}, \ \forall u \in \Gamma(x) \qquad (6.24)$$

and

$$G(x,u) = g(x,u)\tau(x,u), \ \forall x \in \mathcal{X}, \ \forall u \in \Gamma(x) \qquad (6.25)$$

the formulation of Equations 6.18-6.22 can be transformed to the following linear programming formulation, in variables $y^{\Psi}(x,u), x \in \mathcal{X}, \ u \in \Gamma(x)$:

$$V^* = \max_{y^{\Psi}(x,u)} \sum_{x \in \mathcal{X}} \sum_{u \in \Gamma(x)} G(x,u)y^{\Psi}(x,u) \qquad (6.26)$$

s.t.

$$\sum_{u \in \Gamma(x)} y^{\Psi}(x,u) - \sum_{x' \in \mathcal{X}} \sum_{u \in \Gamma(x')} p(x;x',u)y^{\Psi}(x',u) = 0, \ \forall x \in \mathcal{X} \qquad (6.27)$$

$$\sum_{x \in \mathcal{X}} \sum_{u \in \Gamma(x)} \tau(x,u)y^{\Psi}(x,u) = 1 \qquad (6.28)$$

$$y^{\Psi}(x,u) \geq 0, \ \forall x \in \mathcal{X}, \ \forall u \in \Gamma(x) \qquad (6.29)$$

Variables $y^{\Psi}(x,u)$ can be interpreted as the rates according to which the underlying stochastic process finds itself in state x, having selected the control action u at that state, when operated under policy Ψ. Similarly, $\tau(x,u)$ denotes the process expected sojourn time at state x, after having selected control action u, and $G(x,u)$ denotes the corresponding expected gain to be accumulated during that time. The optimal solution, $y^{\Psi*}(x,u)$, of the above formulation, partitions the state space $\mathcal{X}$ into two classes, $\mathcal{X}_r$ and $\mathcal{X}_t$: Class $\mathcal{X}_r$ contains

An algorithm for defining the optimal policy Ψ^* on transient states $x \in \mathcal{X}_t$
Input: A communicating expected average gain CTMDP problem instance, $(\mathcal{X}, \mathcal{U}, \Gamma, \Lambda, p, g)$, together with a solution $y^{\Psi^*}(x, u)$ of the LP formulation of Equations 6.26-6.29
Output: The action selection distribution $\psi^{\Psi^*}(\cdot; x)$ for each transient state $x \in \mathcal{X}_t$

1 $\bar{\mathcal{X}}_r := \{x \in \mathcal{X} : \sum_{u' \in \Gamma(x)} y^{\Psi^*}(x, u') > 0\}$;

2 While $\mathcal{X} \backslash \bar{\mathcal{X}}_r \neq \emptyset$ do

   (a) Find a state $x \in \mathcal{X} \backslash \bar{\mathcal{X}}_r$ and an action $u^* \in \Gamma(x)$ s.t. $\sum_{x' \in \bar{\mathcal{X}}_r} p(x'; x, u^*) > 0$;

   (b) $\forall u \in \Gamma(x)$, if $u = u^*$ then $\psi^{\Psi^*}(u; x) := 1$ else $\psi^{\Psi^*}(u; x) := 0$;

   (c) $\bar{\mathcal{X}}_r := \bar{\mathcal{X}}_r \cup \{x\}$

   endwhile

3 RETURN $\psi^{\Psi^*}(\cdot; x)$ for each $x \in \mathcal{X}_t$.

Figure 6.1.   Defining the optimal policy Ψ^* on transient states $x \in \mathcal{X}_t$

the states $x \in \mathcal{X}$ with $\sum_{u' \in \Gamma(x)} y^{\Psi^*}(x, u') > 0$, that constitute the *recurrent* states during the system operation under policy Ψ^*. Class $\mathcal{X}_t$ contains the states $x \in \mathcal{X}$ with $\sum_{u' \in \Gamma(x)} y^{\Psi^*}(x, u') = 0$, that are the system *transient* states. For the recurrent states $x \in \mathcal{X}_r$, the optimal policy Ψ^* is defined by setting:

$$\psi^{\Psi^*}(u; x) = \frac{y^{\Psi^*}(x, u)\tau(x, u)}{\sum_{u' \in \Gamma(x)} y^{\Psi^*}(x, u')\tau(x, u')}, \quad \forall u \in \Gamma(x) \tag{6.30}$$

For the remaining transient states, the optimal policy is defined by any action selection scheme that establishes a path from each transient state to the class of recurrent states; a detailed algorithm for identifying such a set of paths covering all transient states, is provided in Figure 6.1. Finally, it can be shown that, for any recurrent state $x \in \mathcal{X}_r$, any basic feasible solution of the LP formulation of Equations 6.26-6.29 will have $y^{\Psi^*}(x, u) > 0$ for only one control action $u \in \Gamma(x)$ (Puterman, 1994). Therefore, the optimal policy Ψ^*, returned by this approach, will be *deterministic*.

Bellman's equation and Dynamic Programming.   An alternative solution approach to the considered expected average gain CTMDP problem is provided by the following theorem:

THEOREM 6.2 *(Puterman, 1994; Bertsekas, 1995b) The expected average gain V^* for a communicating CTMDP problem is independent from the process initial state x_0. Furthermore, there exists a vector $h(x)$, $x \in \mathcal{X}$, such that V^* together with h satisfy the following equation for every state $x_i \in \mathcal{X}$:*

$$h^*(x_i) = \max_{u \in \Gamma(x_i)} \left\{ G(x_i, u) - V^* \tau(x_i, u) + \sum_j p(x_j; x_i, u) h^*(x_j) \right\} \quad (6.31)$$

Finally, the deterministic policy Ψ^*, that is defined by setting for all $x_i \in \mathcal{X}$,

$$\psi^{\Psi^*}(u; x_i) = \begin{cases} 1 & \text{for some } u = \arg\max_{u \in \Gamma(x_i)} \{G(x_i, u) - \\ & V^* \tau(x_i, u) + \sum_j p(x_j; x_i, u) h^*(x_j)\} \\ 0 & \text{otherwise} \end{cases} \quad (6.32)$$

is optimal.

Equation 6.31 is known as *Bellman's equation* for the considered CTMDP problem. The parameters $\tau(x, u)$ and $G(x, u)$ that appear in it are respectively defined by Equations 6.24 and 6.25. Furthermore, the vector $h^*(x_i)$ is determined by Equation 6.31 only up to translation by a constant. The difference $h^*(x_i) - h^*(x_j)$ equals the asymptotic relative difference in total reward that results from starting the process in state x_i instead of state x_j, and subsequently operating under an optimal policy Ψ^*; for that reason, the vector $h^*(x)$, $x \in \mathcal{X}$, is characterized as the *optimal relative value function (RVF)* (Puterman, 1994). Finally, for any given V^* and h^*, Equation 6.32 defines the optimal policy Ψ^* by selecting, for each state $x_i \in \mathcal{X}$, a control action $u \in \Gamma(x_i)$ that maximizes the expected gain *differential*

$$\left[G(x_i, u) + \sum_j p(x_j; x_i, u) h^*(x_j) \right] - V^* \tau(x_i, u)$$

to be experienced during the next transition; for that reason, policy Ψ^* is said to be a *"greedy"* policy defined by V^* and h^*.

From an algorithmic standpoint, Equation 6.32 suggests that an optimal policy, Ψ^*, for a communicating expected average gain CTMDP problem, can be obtained by first computing the optimal relative value function, h^*, together with the average gain, V^*, and subsequently identifying a corresponding greedy policy. Indeed, there is an entire class of solution algorithms for the considered MDP problem that have been based on this idea; such algorithms are characterized as *Dynamic Programming (DP)* algorithms, and their relevant theory can be found in (Puterman, 1994; Bertsekas, 1995b).

1.2   Modelling the performance-oriented control problem of logically controlled RAS as an expected average gain CTMDP problem

In this section we provide a detailed formulation of the performance-oriented control problem of logically controlled RAS as an expected average gain CT-MDP problem, under the assumptions that

ASSUMPTION 9 *the considered RAS belongs to the class of DIS-CON-RAS,*

and

ASSUMPTION 10 *the performance objective under consideration is the max-imization of the RAS throughput.*

Restriction of the analysis to the class of DIS-CON-RAS is justified by (i) ex-pository considerations, but also by (ii) the material in Section 3.2 of Chapter 5, which indicates that DIS-CON-RAS have a central role in the deployment of the control function for more general RAS classes. Focusing on the criterion of maximizing the RAS throughput is justified by the fact that, for many practical applications modelled by the RAS class of Definition 1, the most crucial per-formance objective is to *maximize the productivity* of the engaged resources. However, as it will be revealed in the subsequent discussion, the key ideas and techniques underlying the presented formulation can be easily modified to ac-commodate other performance objectives, and therefore, the development of this formulation has also prototypical value. Finally, as we saw in the introduc-tory material of Section 1.1, CTMDP formulations presume that the sojourn times spent by the underlying stochastic process at its various states, are ex-ponentially distributed. Hence, the subsequent discussion will also presume initially that

ASSUMPTION 11 *the distributions D_{jk} introduced in item 5 of Definition 1, are* exponential *with rate $\lambda_{jk} > 0$.*

This assumption will be relaxed in the last part of this section, where it is shown that the proposed methodology can be extended to RAS with more general timing distributions, through the approximation of these distributions by a class of distributions generated by appropriately structured semi-Markov processes.

The MDP formulation

In order to formulate the RAS performance control problem defined by As-sumptions 9-11 as a CTMDP problem, we must provide a detailed character-ization of the various components of the corresponding six-tuple $(\mathcal{X}, \mathcal{U}, \Gamma, \Lambda, p, g)$. The characterization of these components is performed next through a series of steps, that elaborate further the operation of a DIS-CON-RAS that (i) satisfies Assumptions 10 and 11, and (ii) it is under the control of some correct supervisory control policy (SCP) Δ (c.f., Definitions 8 and 9, in Chapter 2, for the characterization of a correct SCP).

Refining the policy-admissible space $S_r(\Delta)$. Consider a RAS $\Phi = < \mathcal{R}, C, \mathcal{P}, \mathcal{A}, \mathcal{T} >$ that satisfies Assumptions 9-11, and it is logically controlled by some correct SCP Δ. Then, the temporal aspects of its behavior can be captured through the following refinement of the FSA model introduced in Section 1.2 of Chapter 2:

1. Each state component $s(q(j, k)) \in S_r(\Delta)$ is expanded to a 3-dimensional vector

$$[s^p(q(j, k)), s^f(q(j, k)), s^b(q(j, k))]$$

   Components $s^p(q(j, k))$ and $s^f(q(j, k))$ indicate the numbers of process instances of stage Ξ_{jk} that are, respectively, in *processing* or *finished*. Component $s^b(q(j, k))$ is a *binary*-valued component, with the value of 1 indicating that the advancement of the finished processes in stage Ξ_{jk} is currently *blocked*. Furthermore, an additional component $s^b(q(j, 0))$ is added for each process type Π_j. This component is also *binary*-valued, and its value of 1 indicates that the loading of any further instances of process type Π_j is currently *blocked*.

2. At any point in time, there should be at least one process instance under processing in the system, i.e., the system operation should be *globally non-idling*. The feasibility of this requirement is guaranteed by (i) the correctness of the applied SCP, and (ii) the further assumption of an *infinite* number of instances from each process type waiting to be loaded in the considered RAS; this last assumption is consistent with the posed objective of throughput maximization.

3. At any RAS state, the completion of an active process instance immediately resets all the state components of $s^b()$ type to 0, i.e., it unblocks all the process loading and advancing events.

4. A state containing unblocked loading and advancing events is followed by a sequence of *immediate* transitions, each of which either (i) advances an unblocked finished process instance to its next stage, or (ii) loads a new process instance of some unblocked process type, or (ii) sets an $s^b()$ state component to its blocked state. Process instances j_j having completed a *"sink"* processing stage Ξ_{jk} in the corresponding process-defining graph $\mathcal{G}_j$ cannot be blocked, but they must be unloaded from the system.[5] The

---

[5]This requirement implies that state components $s^f(q(j, k))$ and $s^b(q(j, k))$ for "sink" processing stages Ξ_{jk} will always be equal to zero, and therefore, they can be dropped from the state vector.

advancement of any other process instance to its next state, or the loading of a new process instance, must be admissible by the applied SCP Δ.

The underlying continuous-time semi-Markov Decision Process.   It is clear from the description of the previous paragraph that states containing unblocked finished process instances or unblocked loading events present zero sojourn times, and therefore, they are characterized as *vanishing*. The remaining states contain only process instances that are in processing or blocked. These states will be characterized as *tangible*, since they have a finite sojourn time, that is determined by the *"exponential race"* of all the process instances that are in processing, for completion of their running processing stage. More specifically, the sojourn times experienced at tangible states are exponentially distributed with a rate equal to $\sum_{(j,k)}[\lambda_{jk} \cdot s^p(q(j,k))]$. Let $S_{rv}(\Delta)$ and $S_{rt}(\Delta)$ denote respectively the sets of vanishing and tangible states that are reachable under the RAS supervision by SCP Δ. Then, the considered RAS performance control problem reduces to *the determination of a probability distribution for each vanishing state, that will regulate the execution of the corresponding Δ-admissible immediate events, so that the resulting system throughput is maximized*. However, the decision process underlying this problem is not exactly CTMDP, since it involves states with, both, exponential and zero sojourn times. In fact, such a decision process is said to be *semi-Markov (semi-MDP)*, in the relevant literature (Puterman, 1994). Next, we further reduce the semi-MDP problem described above, to a CTMDP problem.

The induced CTMDP.   The above semi-MDP formulation of the considered RAS performance-oriented control problem can be further reduced to a CTMDP problem, by focusing on the transitions of the underlying stochastic process among the subclass of vanishing states that are entered from some tangible state, upon the finishing of some RAS process instance. Let us denote this particular subclass of vanishing states by $S_{rv}^0(\Delta)$; i.e., $S_{rv}^0(\Delta)$ defines the *state space* of the induced process. According to item 3 in the description of the refined state space $S_r(\Delta)$, any state $s \in S_{rv}^0(\Delta)$ enables all the Δ-admissible process loading and advancing events. The set of all the *control actions available at state* $s \in S_{rv}^0(\Delta)$ is defined by the set $\Sigma(s)$ of all the Δ-admissible immediate transition sequences σ that can be executed starting from state s; indeed, enabling of the process loading and advancing events in each transition sequence σ is completely at the jurisdiction of the RAS controller. On the other hand, the induced process states $s' \in S_{rv}^0(\Delta)$, that can result from s by taking an action $\sigma \in \Sigma(s)$, are determined by the transitions that are enabled in the tangible state s'' that is obtained in the original semi-MDP from the execution of sequence σ upon state s; i.e., for any given control action $\sigma \in \Sigma(s)$, the resulting states s' correspond to the finishing of some process instance at stage

Ξ_{jk} with $(s'')^p(q(j,k)) > 0$. The corresponding *branching probabilities* are determined by

$$p(s'; s, \sigma) = \frac{\lambda_{jk} \cdot (s'')^p(q(j,k))}{\sum_{(j',k')}[\lambda_{j'k'} \cdot (s'')^p(q(j',k'))]} \quad (6.33)$$

The *sojourn time* associated with the transition resulting from the selection of control action σ at state s is exponentially distributed with *rate*

$$\Lambda(s, \sigma) = \sum_{(j,k)}[\lambda_{jk} \cdot (s'')^p(q(j,k))] \quad (6.34)$$

and *mean value*

$$\tau(s, \sigma) = \frac{1}{\sum_{(j,k)}[\lambda_{jk} \cdot (s'')^p(q(j,k))]} \quad (6.35)$$

The above characterization of the branching probabilities and the state sojourn times indicates clearly that the induced decision process is CTMDP.

Characterizing the objective function of the CTMDP problem. To completely define the considered CTMDP problem, it remains to specify the *gain rate* $g(s, \sigma)$ associated with each state-action pair (s, σ), and the *objective function* under consideration. The characterization of these two issues is determined by our focus on the long-run system throughput as the performance criterion of interest. However, instead of defining the gain rate $g(s, \sigma)$, we proceed to define the *expected immediate gain*, $G(s, \sigma)$, from taking action $\sigma \in \Sigma(s)$ in state s.[6] $G(s, \sigma)$ is naturally defined by the probability that some process instance will be completed and unloaded during the transition resulting from the state-action pair (s, σ), i.e.,

$$G(s, \sigma) = \sum_{(j,k)} \frac{\lambda_{jk} \cdot (s'')^p(q(j,k))}{\sum_{(j',k')}[\lambda_{j'k'} \cdot (s'')^p(q(j',k'))]} \cdot I_{\{\Xi_{jk} \text{ is a "sink" stage of } \mathcal{G}_j\}} \quad (6.36)$$

The long-run system throughput obtained by initializing the RAS at some reachable Δ-admissible state s^0 and subsequently applying some *policy* Ψ for governing the action selection at each state s, is characterized by the *expected average gain* $V^\Psi(s^0)$, which is mathematically defined as follows:

$$V^\Psi(s^0) = \liminf_{t \to \infty} \frac{1}{t} E_{s^0}^\Psi[\sum_{n=0}^{\infty} \int_0^t G(s_n, \sigma_n) \cdot \delta(t - t_n)] \quad (6.37)$$

---

[6]As we saw in the relevant discussion of Section 1.1, while $g(s, \sigma)$ is a primitive concept in the characterization of the CTMDP structure, $G(s, \sigma)$ is the main quantity employed in the LP formulation of Equations 6.26-6.29, as well as in and the definition of the corresponding greedy policy.

In Equation 6.37, t_n denotes the time of the n-th transition in the considered CTMDP, assuming that the RAS operation is started at state s^0 in time $t_0 = 0$, and it is controlled by policy Ψ. The expectation $E_{s^0}^{\Psi}[\cdot]$ is taken over all the corresponding sample paths of the considered CTMDP, and the function $\delta(t)$ is Kronecker's delta function. Our objective is to identify a policy Ψ^* such that

$$V^{\Psi^*}(s) = \sup_{\Psi} \left\{ V^{\Psi}(s), \ \forall s \in S_{rv}^0(\Delta) \right\} \tag{6.38}$$

Solving the induced CTMDP problem.   The observation of the SCP Δ during the construction of the state space of the aforementioned CTMDP guarantees that there exists a deterministic policy that can take the RAS from any given state $s \in S_{rv}^0(\Delta)$ to any other state $s' \in S_{rv}^0(\Delta)$. Hence, the considered CTMDP is classified as *communicating*, and, according to the relevant CTMDP theory presented in Section 1.1, it has a *constant* optimal expected average gain, V^*, for all initial states $s \in S_{rv}^0(\Delta)$, which is provided by some optimal *unichain* policy Ψ^*. Both, V^* and Ψ^* can be computed by solving the following LP formulation in variables $y^*(s, \sigma)$, $s \in S_{rv}^0(\Delta)$, $\sigma \in \Sigma(s)$:

$$\max \sum_{s \in S_{rv}^0(\Delta)} \sum_{\sigma \in \Sigma(s)} G(s, \sigma) y^*(s, \sigma) \tag{6.39}$$

s.t.

$$\sum_{\sigma \in \Sigma(s)} y^*(s, \sigma) - \sum_{s' \in S_{rv}^0(\Delta)} \sum_{\sigma \in \Sigma(s')} p(s; s', \sigma) y^*(s', \sigma) = 0, \ \forall s \in S_{rv}^0(\Delta) \tag{6.40}$$

$$\sum_{s \in S_{rv}^0(\Delta)} \sum_{\sigma \in \Sigma(s)} \tau(s, \sigma) y^*(s, \sigma) = 1 \tag{6.41}$$

$$y^*(s, \sigma) \geq 0, \ \forall s \in S_{rv}^0(\Delta), \forall \sigma \in \Sigma(s) \tag{6.42}$$

Variables $y^*(s, \sigma)$ can be interpreted as the *rates* with which the RAS finds itself in state s, having selected control action σ in that state. The value of the optimal gain, V^*, is equal to the optimal objective value of the above formulation, and a deterministic optimal policy Ψ^* is provided by $y^*(s, \sigma)$ as indicated in Section 1.1.

Incorporating additional throughput-ratio constraints. In case of RAS with more than one process types, the definition of the process gains and the problem objective function provided by Equations 6.36-6.38, implies that the ultimate objective is the *unconditional* maximization of the cumulative throughput across all process types. In most practical situations, the throughput with respect to the various process types will be required to observe some *ratio constraints*, i.e.,

$$TH_j = \rho_j TH_1, \quad \forall j \geq 2 \tag{6.43}$$

The resulting problem is a *constrained* CTMDP problem ((Puterman, 1994), Section 8.9). Equation 6.43 suggests that, in this case, it is pertinent to seek the maximization of TH_1, while observing the ratio constraints. Since we need to differentiate the throughput attained with respect to each process type, the constrained version of the problem will employ a *set* of expected immediate gains, $G_j(s, \sigma)$, $j = 1, \ldots, n$; gain $G_j(s, \sigma)$ corresponds to process type Π_j, and it is defined by

$$G_j(s, \sigma) = \sum_k \frac{\lambda_{jk} \cdot (s'')^p(q(j, k))}{\sum_{(j', k')} [\lambda_{j'k'} \cdot (s'')^p(q(j', k'))]} \cdot I_{\{\Xi_{jk} \text{ is a "sink" stage of } \mathcal{G}_j\}} \tag{6.44}$$

The optimal long-term throughputs, TH_j^*, and an optimal policy Ψ^* for the constrained version of the RAS performance control problem can be obtained by an LP formulation similar to that of Equations 6.39-6.42. The objective function for this new formulation is

$$\max TH_1 \equiv \sum_{s \in S_{rv}^0(\Delta)} \sum_{\sigma \in \Sigma(s)} G_1(s, \sigma) y^*(s, \sigma) \tag{6.45}$$

and the constraint set is augmented with the following constraint, that expresses the throughput ratio requirement of Equation 6.43:

$$\forall j \geq 2, \quad \sum_{s \in S_{rv}^0(\Delta)} \sum_{\sigma \in \Sigma(s)} G_j(s, \sigma) y^*(s, \sigma) - \rho_j \sum_{s \in S_{rv}^0(\Delta)} \sum_{\sigma \in \Sigma(s)} G_1(s, \sigma) y^*(s, \sigma) = 0, \tag{6.46}$$

The optimal throughputs TH_j^*, $j = 2, \ldots, n$, are obtained from TH_1^* by means of Equation 6.43, while a stationary optimal policy Ψ^* can be determined as in the unconstrained case. However, there might *not* exist a deterministic optimal policy for this case.[7]

---

[7] In fact, the policy randomization is the mechanism enabling the satisfaction of the throughput ratio constraint expressed by Equation 6.43.

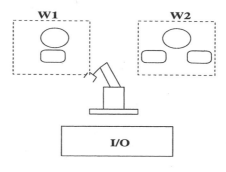

J : W1 -> W2 -> W1

Figure 6.2.   Example: The considered RAS

Table 6.1.   Example: State information for the STD of the considered RAS

s	$s(q(1,1))$	$s(q(1,2))$	$s(q(1,3))$
0	0	0	0
1	1	0	0
2	0	1	0
3	0	0	1
4	1	1	0
5	0	2	0
6	0	1	1

Example.   The following example is adapted from (Choi and Reveliotis, 2003), and it provides a more concrete exposition of the concepts and the techniques discussed above, by applying them for modelling and controlling the resource allocation taking place in the robotic cell depicted in Figure 6.2. This robotic cell consists of two workstations, W_1 and W_2, and a centrally located robot that facilitates the part transfer among the two workstations and the cell I/O port. Each workstation comprises a single processor, and a buffer providing a number of slots; in particular, the buffer of workstation W_1 has a single slot, while the buffer of workstation W_2 has two slots. While visiting any of the two workstations, a part is initially accommodated to one of its buffer slots; at some point, it enters the workstation server for processing, and upon completion, it returns to its allocated slot, where it waits to be transferred to its next stage. Notice that, under the described operational regime, the part holds its allocated buffer slot throughout its entire sojourn in the considered workstation. In the depicted operational mode, the cell supports the production of a single part type,

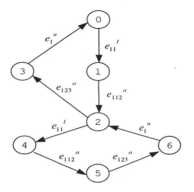

Figure 6.3. Example: The reachable and safe subspace, $S_{rs} \equiv S_r(\Delta^*)$, of the considered RAS

J, that follows the processing route annotated in Figure 6.2. Finally, it is further assumed that the processing times associated with the various processing stages are exponentially distributed, and that the time required for transferring any part between the cell workstations and/or the I/O port, is negligible compared to the experienced processing times.

In order to analyze the cell of Figure 6.2 with respect to the logical control problem of establishing non-blocking supervision for the underlying resource allocation, we can abstract it as a LIN-SU-RAS, Φ, consisting of two resources, R_1 and R_2, corresponding to the buffers of the two workstations, W_1 and W_2, and a single process type Π_1, involving three linearly ordered processing stages, Ξ_{1k}, $k = 1, 2, 3$. The capacities of R_1 and R_2 are $C_1 = 1$ and $C_2 = 2$, while the resource allocation requests posed by the three processing stages, Ξ_{1k}, $k = 1, 2, 3$, are, respectively, $A_{11} = (1, 0)^T$, $A_{12} = (0, 1)^T$ and $A_{13} = (1, 0)^T$. Then, it is easy to see that the optimal non-blocking SCP, Δ^*, for this simple RAS, is expressed by the following linear constraint on the RAS state, s:

$$(1\ 1\ 0) \cdot s \leq 2 \tag{6.47}$$

The resulting reachable and safe subspace of RAS Φ is depicted in Figure 6.3, while the complete characterization of the states depicted in Figure 6.3 is provided in Table 6.1.

In order to characterize the timed dynamics of the resource allocation underlying the operation of the cell of Figure 6.2, we must explicitly model the three different phases characterizing the part sojourn at any visited workstation – i.e., waiting for processing, being processed, and being finished and waiting for transferring to the next stage – and also, the time spend by a part in transport.

Table 6.2. Example: State information for the semi-Markov Decision Process characterizing the timed dynamics of the considered RAS

s	$s^b(q(1,0))$	$<s^p\ s^J\ s^b>(q(1,1))$	$<s^p\ s^J\ s^b>(q(1,2))$	$s^p(q(1,3))$
0	0	0 0 0	0 0 0	0
1	0	1 0 0	0 0 0	0
2	1	1 0 0	0 0 0	0
3	0	0 1 0	0 0 0	0
4	0	0 0 0	1 0 0	0
5	0	1 0 0	1 0 0	0
6	1	0 0 0	1 0 0	0
7	1	1 0 0	1 0 0	0
8	0	0 0 0	0 1 0	0
9	0	0 1 0	1 0 0	0
10	0	1 0 0	0 1 0	0
11	0	0 0 0	0 0 0	1
12	1	0 1 1	1 0 0	0
13	0	0 0 0	2 0 0	0
14	1	1 0 0	0 1 1	0
15	1	0 0 0	0 0 0	1
16	0	0 1 0	0 1 0	0
17	1	0 0 0	2 0 0	0
18	0	0 0 0	1 1 0	0
19	0	0 0 0	1 0 0	1
20	1	0 0 0	1 1 1	0
21	1	0 0 0	1 0 0	1
22	0	0 0 0	0 2 0	0
23	0	0 0 0	0 1 0	1
24	1	0 0 0	0 1 1	1

The most straightforward way to perform this modelling is by redefining the resource set, $\mathcal{R}$, and the process type, Π_1, of the RAS Φ that was employed in the logical analysis of the considered cell, so that they take into account all the aforementioned elements. However, this suggested refinement can be simplified, in the context of the considered example, by taking into consideration the following two observations:

1 The earlier introduced assumption that part transfers among the cell workstations and the I/O station have a duration that is negligible when compared to the expected stage processing times, implies that parts can be presumed to be *immediately* transferred to their next stage, upon the authorization of this transfer, and there is not need to model explicitly stages corresponding to part transferring.

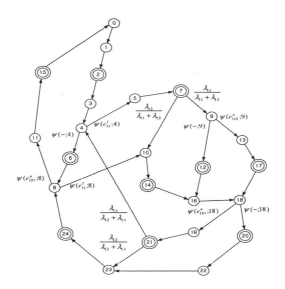

Figure 6.4. Example: The embedded Markov Chain of the Semi-Markov Decision Process characterizing the timed dynamics of the considered RAS

2 The unit buffering capacity of workstation W_1, when combined with the previous description of the part sojourn at any of the cell stations, imply that any part visiting this station will never need to wait for allocation of the station server, but it can proceed directly to processing. A similar effect can be established for workstation W_2, by noticing that the combination of Equation 6.47 with the assumption of immediate part transfers, further imply that there is no advantage in authorizing the transfer of a part from workstation W_1 to workstation W_2, unless the part can immediately start its processing in W_2.

The above two observations subsequently imply that, in order to characterize the timed dynamics of the resource allocation taking place in the robotic cell of Figure 6.2, under the control of the maximally permissive SCP, Δ^*, of Equation 6.47, there is no need to augment the underlying RAS Φ, that was employed for the logical analysis of the considered cell, with any additional resources and stages; the only refinement needed is the state re-definition of RAS Φ along the lines discussed in the earlier theoretical developments of this section. The *embedded* Markov chain of the resulting semi-Markov decision process (semi-MDP) is depicted in Figure 6.4, while the depicted states are further characterized in Table 6.2. As explained in the theoretical discussion of this section, the states of the semi-MDP of Figure 6.4 are partitioned into two classes:

Tangible states, depicted by double-circled nodes in Figure 6.4. Each of these states corresponds to a RAS status where all processes loaded in the system are in processing or blocked. Furthermore, any process-loading events are blocked, and the only enabled events are defined by the termination of the processes which are currently in processing. Hence, transitions out of these states are defined by the events characterizing the completion of a process instance belonging to one of the active processing stages; the corresponding branching probabilities can be obtained as depicted in Figure 6.4, by taking into consideration the exponential nature of the RAS processing times.

Vanishing states, depicted by single-circled nodes in Figure 6.4. These states contain process instances that have finished their running processing stage and request advancement to their next stage, and also, unblocked process-loading events. Hence, these states necessitate some *scheduling decisions* regarding the loading, advancement and/or the blocking of these processes. In the proposed CTMDP-based modelling framework, these decisions are regulated by the action-selection probability distributions, $\psi^{\Psi}(\cdot; s)$, that define the applied stationary randomized policy Ψ. More specifically, each distribution $\psi^{\Psi}(\cdot; s)$ corresponds to a single vanishing state $s \in S_{rv}(\Delta^*)$, and its support is defined by the set of the Δ^*-admissible immediate events at state s; in Figure 6.4, the Δ^*-admissible control actions authorizing the process loading and advancing events are characterized by the corresponding event e, while decisions that block any further progress are indicated by '-'. An additional implicit assumption underlying the development of the semi-Markov decision process depicted in Figure 6.4 and Table 6.2 is that process instances are unloaded immediately from the considered RAS, upon the completion of processing stage Ξ_{13}. Finally, under a well-defined and computationally efficient control policy, the time necessary for making a scheduling decision at any vanishing state, $s \in S_{rv}(\Delta^*)$, will be negligible compared to the processing times associated with the execution of the various processing stages; therefore, for the purposes of the considered analysis, these states are assumed to present zero sojourn time, which explains the characterizations of "tangible" and "vanishing" states.

Figure 6.5 depicts the continuous-time Markov decision process (CTMDP) obtained from the semi-MDP of Figure 6.4, by (i) focusing on the set of vanishing states $S_{rv}^0(\Delta^*)$, that result from tangible states upon the completion some running process instance, (ii) enumerating all the Δ^*-admissible sequences of immediate transitions emanating from these states, and (iii) clustering each of these sequences into a single control action, characterized by the involved process-loading, advancing and/or blocking events. In a strict, formal sense,

Table 6.3. Example: Vanishing state information for the CT-Markov Decision Process of the considered RAS

s_k	$s^b(q(1,0))$	$< s^p\ s^J\ s^b > (q(1,1))$	$< s^p\ s^J\ s^b > (q(1,2))$	$s^p(q(1,3))$
0	0	0 0 0	0 0 0	0
3	0	0 1 0	0 0 0	0
4	0	0 0 0	1 0 0	0
8	0	0 0 0	0 1 0	0
9	0	0 1 0	1 0 0	0
10	0	1 0 0	0 1 0	0
16	0	0 1 0	0 1 0	0
18	0	0 0 0	1 1 0	0
22	0	0 0 0	0 2 0	0
23	0	0 0 0	0 1 0	1

Table 6.4. Example: Tangible state information for the CT-Markov Decision Process of the considered RAS

s_k	$s^b(q(1,0))$	$< s^p\ s^J\ s^b > (q(1,1))$	$< s^p\ s^J\ s^b > (q(1,2))$	$s^p(q(1,3))$
2	1	1 0 0	0 0 0	0
6	1	0 0 0	1 0 0	0
7	1	1 0 0	1 0 0	0
12	1	0 1 1	1 0 0	0
14	1	1 0 0	0 1 1	0
15	1	0 0 0	0 0 0	1
17	1	0 0 0	2 0 0	0
20	1	0 0 0	1 1 1	0
21	1	0 0 0	1 0 0	1
24	1	0 0 0	0 1 1	1

the state space of the aforementioned CTMDP consists of the vanishing states $s \in S_{rv}^0(\Delta^*)$. However, we opted to include in Figure 6.5 the tangible states, s'', resulting from the execution of any control action u in some state $s \in S_{rv}^0(\Delta^*)$, in order to provide a more detailed exposition of the mechanism determining the transition of the considered CTMDP among its various states; complete descriptions of the depicted vanishing and tangible states are provided respectively in Tables 6.3 and 6.4.

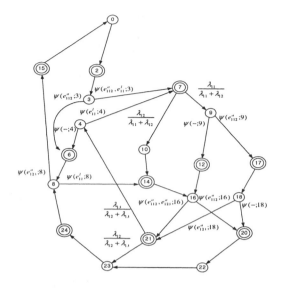

Figure 6.5.   Example: The induced CT-Markov Decision Process for the considered RAS

Next, we characterize the optimal policy Ψ^* for the considered RAS Φ. Since Φ contains only the single process type Π_1, its performance control problem can be formulated as an *unconstrained* expected average gain CTMDP problem, and therefore, there will exist a *deterministic* optimal scheduling policy Ψ^*. Such an optimal policy can be identified, for any particular RAS configuration resulting by the pricing of processing rates λ_{1k}, $k = 1, 2, 3$, through the solution of the corresponding LP formulation defined by Equations 6.39-6.42. In the following, we provide a more general characterization of the structure of the optimal policy, Ψ^*. This characterization is obtained by partially enumerating all the deterministic policies, Ψ, that can be defined for the CTMDP of Figure 6.5, and comparing the resulting throughputs. We start with the following observations:

1 First, notice that for vanishing states 9 and 18, the respective optimal decisions are those corresponding to transitions $9 \rightarrow 17$ and $18 \rightarrow 21$, since the alternative decisions simply introduce pure delays in the system operation.

2 Furthermore, the optimal decisions for states 3 and 4 should lead to the same tangible state; this results from Equation 6.32, and the role of the tangible states, s'', in defining the expected sojourn times and branching probabilities for the various state-action pairs, (s, u).

Table 6.5.   Example: The throughput functions corresponding to the four candidate scheduling policies for the considered RAS

ψ_1	ψ_2	$TH(\psi_1, \psi_2; \lambda_{11}, \lambda_{12}, \lambda_{13})$
0	0	$\dfrac{\lambda_{11}\lambda_{12}\lambda_{13}}{\lambda_{11}\lambda_{12}+\lambda_{12}\lambda_{13}+\lambda_{11}\lambda_{13}}$
0	1	$\dfrac{\lambda_{11}\lambda_{12}\lambda_{13}(\lambda_{12}+\lambda_{13})}{\lambda_{11}\lambda_{13}^2+\lambda_{11}\lambda_{12}\lambda_{13}+\lambda_{11}\lambda_{12}^2+\lambda_{12}^2\lambda_{13}+\lambda_{12}\lambda_{13}^2}$
1	0	$\dfrac{\lambda_{11}\lambda_{12}\lambda_{13}(\lambda_{11}+\lambda_{12})(2\lambda_{12}+\lambda_{13})}{\lambda_{11}^2\lambda_{13}^2+2\lambda_{11}\lambda_{12}^3+\lambda_{11}\lambda_{12}\lambda_{13}^2+2\lambda_{11}^2\lambda_{12}\lambda_{13}+2\lambda_{12}^3\lambda_{13}+\lambda_{12}^2\lambda_{13}^2+2\lambda_{11}^2\lambda_{12}^2+3\lambda_{11}\lambda_{12}^2\lambda_{13}}$
1	1	$\dfrac{\lambda_{11}\lambda_{12}\lambda_{13}(\lambda_{11}+\lambda_{12})(\lambda_{12}+\lambda_{13})}{\lambda_{11}^2\lambda_{13}^2+2\lambda_{11}\lambda_{12}^2\lambda_{13}+\lambda_{11}\lambda_{12}^3+\lambda_{11}\lambda_{12}\lambda_{13}^2+\lambda_{11}^3\lambda_{12}\lambda_{13}+\lambda_{12}^3\lambda_{13}+\lambda_{12}^2\lambda_{13}^2+\lambda_{11}^2\lambda_{12}^2}$

Hence, the optimal policy Ψ^* can be characterized by means of two *binary* variables, $\psi_1 \equiv \psi(e_{112}^a, e_{11}^l; 3)$ and $\psi_2 \equiv \psi(e_{11}^l; 8)$. The formulae providing the RAS throughput that results for each of the four possible pricings of variables ψ_1 and ψ_2, as a function of the processing rates λ_{1k}, $k = 1, 2, 3$, are presented in Table 6.5. Taking the pairwise differences defined over the set of these throughput functions, it can be shown that they satisfy the dominance relationships expressed by the lattice of Figure 6.6, for all positive triplets $< \lambda_{11}, \lambda_{12}, \lambda_{13} >$. Hence, it can be concluded that the optimal deterministic policy Ψ^* is defined by the pair $(\psi_1 = 1, \psi_2 = 1)$.

Closing the discussion of this example, we invite the reader to show that the optimal policy Ψ^* essentially implements a *"First-Buffer-First-Serve (FBFS)"* resource allocation scheme, while abiding to the requirements of the applied SCP Δ^*. On the other hand, the sub-optimal policy Ψ defined by the pair $(\psi_1 = 1, \psi_2 = 0)$, essentially is equivalent to a *"Last-Buffer-First-Serve (LBFS)"* resource allocation scheme, that is also abiding to the requirements of the applied SCP Δ^*. This realization becomes more interesting when viewed in the light of the results presented in (Lu and Kumar, 1991), which have established that, when $C_i \to \infty$, $i = 1, 2$, both policies, FBFS and LBFS, are throughput-optimal. Hence, this example corroborates the remark made in the introductory chapter, that the structure of the optimal scheduling policy obtained when considering the finiteness of system resources other than its processing capacity, and the requirements of the necessary non-blocking SCP, might be drastically different from the structure of the optimal scheduling policies obtained by any analysis that ignores the aforementioned system elements.

Performance analysis and control of logically controlled RAS with non-Markovian behavior. When the considered logically controlled RAS, Φ, involves processing time distributions, D_{jk}, other than exponential, the stochastic process induced on the state space $S_r(\Delta)$ by the adoption of any stationary ran-

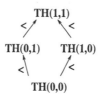

Figure 6.6. Example: The dominance relationships satisfied by the throughput functions, $TH(\psi_1, \psi_2)$, $\psi_1, \psi_2 \in \{0, 1\}$, corresponding to the candidate scheduling policies

domized scheduling policy Ψ, is not semi-Markov anymore, but *generalized semi-Markov*: indeed, the basic structure of the RAS behavior is still characterized by the refinement of the reachable and Δ-admissible state space, $S_r(\Delta)$, introduced in the previous section, but the transition of the underlying stochastic process from any tangible state $s \in S_{rt}(\Delta)$ is determined by the *residual* processing times of the various process instances that are active in that state. As it was mentioned in the discussion of Stochastic Timed Automata, in Section 1.1, GSMP behavior is too complex to be analytically tractable, and thus, simulation is used as the primary means of analysis of the system performance. An introductory discussion on the role of simulation as a performance analysis tool, together with additional references on this topic, can be found in (Cassandras and Lafortune, 1999). Next we briefly discuss an alternative approach to the performance analysis and control of logically controlled RAS with general processing time distributions, D_{jk}, that seeks to reduce the problem to an expected average gain CTMDP problem, by approximating each distribution D_{jk} through a distribution generated by an appropriately structured semi-Markov process. Distributions generated by such semi-Markov processes are known as *Phase-type* distributions, while the resulting approximation method is known as the *method of stages*.

According to the method of stages, each processing stage Ξ_{jk} with non-exponentially distributed processing times, must have its node in the corresponding process-defining graph, $\mathcal{G}_j$, replaced by the state space of a semi-Markov process possessing the structure depicted in Figure 6.7. In Figure 6.7, parameters q_{ij} denote the branching probabilities of the depicted semi-Markov process, and λ_i, $i = 1, \ldots, k$, is the rate of the exponential distribution characterizing the process sojourn time at state i; states 0 and $k + 1$ have zero sojourn times. From states $0, 1, \ldots, k$, the process can transition to any other state $i \in \{1, \ldots, k + 1\}$; however, state $k + 1$ is an absorbing state. The parameters k, q_{ij} and λ_i are selected in a manner that the time required for the process "absorption" to state $k + 1$, when initialized to state 0, has the same statistics with the processing times, t_{jk}, corresponding to processing stage Ξ_{jk}.

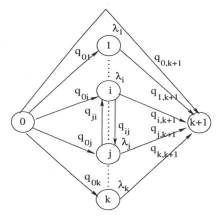

Figure 6.7. The semi-Markov process generating a Phase-type distribution with k phases

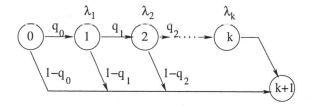

Figure 6.8. The semi-Markov process generating a Coxian distribution with k stages

The feasibility of such an approximation scheme for most distributions D_{jk} of practical interest has been theoretically established by Cox in (Cox, 1955), where he has shown that any distribution having a rational Laplace-Stieltjes transform can be represented *exactly* by a phase-type distribution generated by a semi-Markov process that has the particular structure depicted in Figure 6.8; such a distribution is known as a $k-stage$ *Coxian distribution*. The particular sub-class of Coxian distributions obtained by setting $q_0 = q_1 = \ldots = q_{k-1} = 1$, is known as the class of $k-stage$ *Erlang distributions*; Erlang distributions are extensively used for the approximation of general distributions with coefficient of variation less that one. Another interesting class of phase-type distributions is that of $k-stage$ *hyper-exponential distributions*. These distributions are obtained from the general topology of Figure 6.7 by setting $q_{0,k+1} := 0$; $q_{i,k+1} := 1$, $\forall i \in \{1, \ldots, k\}$, and they are appropriate for approximating general distributions with coefficient of variation greater than one. Notice that both classes of Erlang and hyper-exponential distributions contain the exponential distribution as the special case of $k = 1$.

Clearly, the behavior generated by any single instance of process type Π_j, $j = 1, \ldots, n$, after the substitution of each stage, Ξ_{jk}, possessing non-exponentially distributed processing times, with the state space of an appropriately configured phase-type distribution, corresponds to a semi-Markov process. Hence, the RAS, Φ', resulting from the aforementioned substitutions, is amenable to performance analysis and control through the methodology developed in the earlier parts of this chapter. It must be noticed, however, that, in general, very high levels of approximating accuracy for the target distributions imply state space topologies with a very large number of states, for the corresponding phase-type distributions. Hence, at some point, the modeler must deal with the issue of compromising approximating accuracy for computational tractability of the resulting RAS model. Some typical techniques employed for the management of this dilemma can be found in (Papadopoulos et al., 1993) and the references cited therein. An additional problem arises during the implementation of any controller that has been developed through the approximation method of stages, due to the fact that it is not possible to track the states of the various phase-type distributions, while monitoring the operation of the actual system, since these states are fictitious constructs. Acknowledging this practical constraint, one can seek to identify an optimized scheduling policy, Ψ^*, among the class of policies that exert a control action only upon the completion of the running processing stages of the active process instances.[8] The restriction of the corresponding CTMDP formulation on this class of scheduling policies can be imposed through the appropriate specification of the control action sets, $\Sigma(s)$, of the semi-Markov decision process modelling the behavior of the modified RAS Φ', associated with the various vanishing states, $s \in S'_{rv}(\Delta)$; the relevant implementational details are left to the reader.

2.   The thinning problem revisited

We remind the reader that the *"thinning"* problem was originally introduced in Section 3.2 of Chapter 5, and it was motivated by the need to establish computational tractability for the DIS-CON-RAS-based representation of RAS with more complex process behaviors. More specifically, it was observed in Chapter 5 that the size of the state-machine nets encoding in the DIS-CON-RAS-based representation the target behavior of the considered RAS processes, will be super-polynomially related to the size of the resource-augmented process subnets that constitute the original process representation, since these state-machine nets are essentially defined by the reachable and safe space of the

---

[8]Essentially, this is the price that must be paid for not including the *residual* processing times as part of the refined RAS state. On the other hand, such an inclusion would convert the underlying stochastic process to a *hybrid* system, and it would render intractable the analytical characterization and optimization of its behavior.

corresponding resource-augmented process subnets. In the particular case that the size of any of these induced state-machine nets is deemed to be computationally prohibitive for the effective design and deployment of the sought control function, one can try to restrict the execution of the corresponding process in a subnet of the originally induced net, selected in a way that (i) preserves the safety of the process execution, and (ii) remains within an externally specified size-range. An additional, naturally arising concern is that the selected subnets lead to the best possible performance of the controlled RAS, over the range of performances allowed by the selection constraints. Hence, the thinning problem is an *optimization* problem where the main decision variables concern the structural configuration of the underlying RAS and the design of its logical control function, but the evaluation of these decisions is based on performance-oriented considerations. A systematic analytical formulation of this problem needs to employ concepts and results coming from, both, the logical and the performance-oriented RAS modelling frameworks developed in earlier parts of this book. This section provides such a formulation for the thinning problem. The development of this formulation completes the problem coverage, by providing a thorough analytical characterization of it, but it is also intended as a prototypical example regarding the ability of the CTMDP-based RAS-modelling framework of Section 1.2 to provide analytical characterizations for RAS configuration design problems that couple logical and performance-oriented control issues.

An analytical characterization of the thinning problem. Based on the previous discussion, the thinning problem can be abstracted as the selection of a subset of edges, $\mathcal{E}_j^S$, from each process graph, $\mathcal{G}_j$, of a DIS-CON-RAS, Φ, such that (i) $|\mathcal{E}_j^S| = \mu_j$, $\forall j$, (ii) and some performance index of the RAS Φ', that results by restricting each process type Π_j to its process plans encompassed by $\mathcal{E}_j^S$, is maximized. Parameters μ_j are externally specified, and they quantify the size constraints imposed on the selected process subnets. Next we provide a detailed, analytical characterization of the thinning problem for the case that the observed performance objective is the maximization of the RAS throughput, which was also the performance objective considered in the development of Section 1.2.

Consider the graph $\mathcal{G}_j$ characterizing the execution logic of some process type Π_j of a DIS-CON-RAS Φ, and a selection of edges from this graph, $\mathcal{E}_j^S \subseteq \mathcal{E}_j$. The subgraph $\mathcal{G}_j(\mathcal{E}_j^S)$ that encodes all the process plans for process type Π_j involving only transitions corresponding to edges in $\mathcal{E}_j^S$, can be computed by the algorithm of Figure 6.9. Then, given a DIS-CON-RAS Φ and sets $\mathcal{E}_j^S \subseteq \mathcal{E}_j$ for each process type Π_j, let $\Phi(\mathcal{E}_1^S, \ldots, \mathcal{E}_n^S)$ denote the DIS-CON-RAS obtained by substituting each process graph $\mathcal{G}_j$ by $\mathcal{G}_j(\mathcal{E}_j^S)$, in the definition of Φ. Of particular interest for the pursued formulation are all the

An algorithm for characterizing all the process plans encompassed in an edge selection $\mathcal{E}_j^S \subseteq \mathcal{E}_j$ of the process graph $\mathcal{G}_j$ corresponding to a process type Π_j of a DIS-CON-RAS Φ

Input: A process graph $\mathcal{G}_j$ corresponding to a process type Π_j of a DIS-CON-RAS Φ, and a subset of its edge set, $\mathcal{E}_j^S \subseteq \mathcal{E}_j$

Output: The graph $\mathcal{G}_j(\mathcal{E}_j^S)$ that encodes all the process plans of process type Π_j encompassed in $\mathcal{E}_j^S$

1 Construct the graph $\bar{\mathcal{G}}_j$, by adding to graph $\mathcal{G}_j$ a node v_{j0} together with the set of edges $\{(v_{j0}, v_{jk}) : v_{jk} \in V_j^{\nearrow}\} \cup \{(v_{jk}, v_{j0}) : v_{jk} \in V_j^{\searrow}\}$.

2 Construct the subgraph $\bar{\mathcal{G}}_j(\mathcal{E}_j^S)$ of $\bar{\mathcal{G}}_j$, with node set $\{v_{jk}, \ k = 0, 1, \ldots, l(j)\}$, and edge set $\mathcal{E}_j^S \cup \{(v_{j0}, v_{jk}) : v_{jk} \in V_j^{\nearrow}\} \cup \{(v_{jk}, v_{j0}) : v_{jk} \in V_j^{\searrow}\}$.

3 Trim the graph $\bar{\mathcal{G}}_j(\mathcal{E}_j^S)$ with respect to node v_{j0}, to obtain the graph $\bar{\mathcal{G}}_j^{Tr}(\mathcal{E}_j^S)$.

4 Delete the node v_{j0} and its incident edges from $\bar{\mathcal{G}}_j^{Tr}(\mathcal{E}_j^S)$, to obtain the graph $\mathcal{G}_j(\mathcal{E}_j^S)$.

5 RETURN $\mathcal{G}_j(\mathcal{E}_j^S)$.

Figure 6.9.   Characterizing all the process plans encompassed in an edge selection $\mathcal{E}_j^S \subseteq \mathcal{E}_j$ of the process graph $\mathcal{G}_j$ corresponding to a process type Π_j of a DIS-CON-RAS Φ

DIS-CON-RAS $\Phi(\mathcal{E}_1^S, \ldots, \mathcal{E}_n^S)$ for which (i) the edge sets $\mathcal{E}_j^S$ have $|\mathcal{E}_j^S| = \mu_j$, where $\mu_j \leq |\mathcal{E}_j|$ is externally defined, and (ii) $\mathcal{G}_j(\mathcal{E}_j^S) \neq NULL$, $\forall j$. In the following, the distinct edge selections defining these DIS-CON-RAS will be denoted by $\{\mathcal{S}_1, \ldots, \mathcal{S}_N\}$, and the corresponding DIS-CON-RAS by $\Phi(\mathcal{S}_u)$, $u = 1, \ldots, N$. Clearly, N is finite; in particular, $N \leq \prod_{j=1}^n \binom{|\mathcal{E}_j|}{\mu_j}$.

In the context of this formalism, the thinning problem can be posed as follows: Given a DIS-CON-RAS Φ, a correct LES Δ, target throughput ratios ρ_j for $j = 2, \ldots, |\mathcal{P}|,$[9] and parameters $\mu_j \leq |\mathcal{E}_j|$, $j = 1, \ldots, |\mathcal{P}|$, find an edge selection $\mathcal{S}^*$ such that

$$\mathcal{S}^* = \arg\max_{\mathcal{S}} \left\{ TH_1^*(\Phi(\mathcal{S}); \Delta, \rho_2, \ldots, \rho_{|\mathcal{P}|}) \right\} \qquad (6.48)$$

---

[9]In the case of DIS-CON-RAS with a single process type, this data set is vacuous, and the unconstrained version of the throughput maximization problem must be employed.

s.t.

$$|\mathcal{E}_j^S| = \mu_j \wedge \mathcal{G}_j(\mathcal{E}_j^S) \neq NULL, \ \forall \mathcal{E}_j^S \in \mathcal{S} \tag{6.49}$$

The maximal throughput, $TH_1^*(\Phi(\mathcal{S}); \Delta, \rho_2, \ldots, \rho_{|\mathcal{P}|})$, for DIS-CON-RAS $\Phi(\mathcal{S})$, can be computed through the techniques presented in Section 1.2. In the particular case that all the timing distributions, D_{jk}, of the considered RAS, Φ, are exponential, the maximal throughput $TH_1^*(\Phi(\mathcal{S}); \Delta, \rho_2, \ldots \rho_{|\mathcal{P}|})$ can be computed through the formulation of Equations 6.45, 6.40–6.42 and 6.46, by forcing to zero every variable $y^*(s, \sigma)$ that corresponds to a transition leading to a state s' which does not belong to $S_r(\Phi(\mathcal{S}); \Delta)$, the reachable Δ-admissible subspace for the DIS-CON-RAS $\Phi(\mathcal{S})$.

Next we establish a dominance relationship that will allow the further restriction of the set of selections to be considered in the solution of the thinning problem defined by Equations 6.48 and 6.49. Given two selections $\mathcal{S}_1$ and $\mathcal{S}_2$ satisfying Constraint 6.49, let $\mathcal{S}_1 \preceq \mathcal{S}_2$ denote that graph $\mathcal{G}(\mathcal{E}_j^{S_1})$ is a subgraph of $\mathcal{G}(\mathcal{E}_j^{S_2})$, for all j. The relationship '$\mathcal{S}_1 \preceq \mathcal{S}_2$' defines a *partial order* on the set of selections satisfying Constraint 6.49, and we shall say that selection $\mathcal{S}_1$ is *dominated by* – or it is *smaller than* – selection $\mathcal{S}_2$. Clearly, $\mathcal{S}_1 \preceq \mathcal{S}_2$ implies that $S_r(\Phi(\mathcal{S}_1); \Delta) \subseteq S_r(\Phi(\mathcal{S}_2); \Delta) \subseteq S_r(\Phi; \Delta)$. Suppose that the evaluation of the optimal throughput, $TH_1^*(\Phi(\mathcal{S}_i); \Delta, \rho_2, \ldots, \rho_{|\mathcal{P}|})$, for RAS $\Phi(\mathcal{S}_i)$, $i = 1, 2$, is performed through the method of stages. Then, the above set inclusions imply that, (i) every feasible solution for the LP formulation estimating $TH_1^*(\Phi(\mathcal{S}_1); \Delta, \rho_2, \ldots, \rho_{|\mathcal{P}|})$ is also a feasible solution for the LP formulation estimating $TH_1^*(\Phi(\mathcal{S}_2); \Delta, \rho_2, \ldots, \rho_{|\mathcal{P}|})$, while (ii) the objective function of the former formulation is subsumed in the objective function of the latter. The last two observations further imply that $TH_1^*(\Phi(\mathcal{S}_1); \Delta, \rho_2, \ldots, \rho_{|\mathcal{P}|}) \leq TH_1^*(\Phi(\mathcal{S}_2); \Delta, \rho_2, \ldots, \rho_{|\mathcal{P}|})$. Hence, it can be concluded that, among the selections that satisfy Constraint 6.49, only those that are *maximal* with respect to relationship '$\preceq$' need to be considered in the solution of Equation 6.48.

Example. To provide a more concrete demonstration of the concepts underlying the formulation of Equations 6.48 and 6.49, consider the DIS-CON-RAS, Φ, of Figure 6.10. This RAS consists of three resources R_1, R_2 and R_3, each possessing unit capacity, i.e., $C_1 = C_2 = C_3 = 1$. In its current configuration, the system supports a single process type, Π_1, with the sequential logic characterized by graph $\mathcal{G}_1$; specifically, there are two possible process plans for this process type: $\Xi_{11} \rightarrow \Xi_{13}$ and $\Xi_{12} \rightarrow \Xi_{13}$. Figure 6.10 indicates also the resource allocation requests A_{jk} associated with each processing stage Ξ_{jk}: stage Ξ_{11} requires one unit of resource R_1 to support its execution, stage Ξ_{12} requires one unit of resource R_2, and stage Ξ_{13} requires one unit of resource R_3. Figure 6.11 depicts the STD structure for the reachable subspace, S_r, of

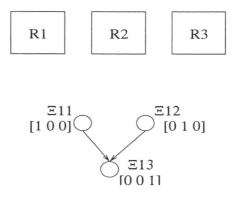

Figure 6.10. Example: The considered DIS-CON-RAS

Table 6.6. State information for the STD of the considered DIS-CON-RAS

s_k	$s(q(1,1))$	$s(q(1,2))$	$s(q(1,3))$
0	0	0	0
1	1	0	0
2	0	1	0
3	1	1	0
4	0	0	1
5	0	1	1
6	1	0	1
7	1	1	1

Φ, while the detailed characterization of the RAS states corresponding to the various STD nodes is provided in Table 6.6. Notice that, for the considered RAS, $S_r = S_{rs}$, which is the result of the single-process structure of Φ and the acyclic structure of graph $\mathcal{G}_1$. Hence, there is no need for any externally imposed SCP Δ; we shall denote this effect by setting $\Delta = \Im$. To complete the characterization of RAS Φ, suppose also that the processing times, t_{1k}, for the three stages Ξ_{1k}, $k = 1, 2, 3$, are exponentially distributed with corresponding rates λ_k.

Next, consider that the "thinning" requirement of $\mu_1 = 1$ is imposed on the operation of the aforementioned RAS. It is obvious from the structure of the graph $\mathcal{G}_1$, depicted in Figure 6.10, that this thinning requirement essentially

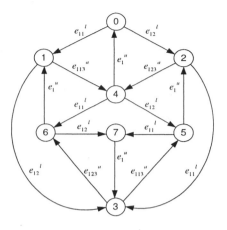

Figure 6.11. The State Transition Diagram (STD) for the DIS-CON-RAS of Figure 6.10

implies the enactment of *only* one of the two possible process plans during the system operation. Technically, these two selections are defined by the edge sets $\mathcal{E}_1^{S_1} = \{(\Xi_{11}, \Xi_{13})\}$ and $\mathcal{E}_1^{S_2} = \{(\Xi_{12}, \Xi_{13})\}$, both of which are maximal with respect to relation '$\preceq$'. The subspaces $S_r(\Phi(\mathcal{S}_i); \Im)$, $i = 1, 2$, can be easily obtained from the information provided in the state space S_r, of Figure 6.11 and in Table 6.6. For this simple case, once the subspaces $S_r(\Phi(\mathcal{S}_i); \Im)$ have been developed, it is also possible to obtain closed-form expressions for the optimal throughputs $TH^*(\Phi(\mathcal{S}_i); \Im)$, $i = 1, 2$, through the steady-state analysis of the induced CTMC's describing the restriction of the RAS operation on these subspaces. Following this workplan, the reader is invited to show that the optimal selection, $\mathcal{S}^*$, for the considered thinning problem, is given by

$$\mathcal{S}^* = \begin{cases} \mathcal{S}_1, & \text{for } \lambda_1 \geq \lambda_2 \\ \mathcal{S}_2, & \text{for } \lambda_1 \leq \lambda_2 \end{cases}$$

◇

A MIP formulation of the thinning problem for Markovian RAS. In the case of RAS with exponentially distributed stage processing times, Equation 6.41 implies that, in the computation of $TH_1^*(\Phi(\mathcal{S}); \Delta, \rho_2, \ldots, \rho_{|\mathcal{P}|})$ for some given edge selection $\mathcal{S}$, the requirement of forcing to zero every variable $y^*(s, \sigma)$ that corresponds to a transition leading to a state $s' \notin S_r(\Phi(\mathcal{S}); \Delta)$, can be expressed by the addition of the following constraint in the LP formulation of Equations 6.45, 6.40–6.42 and 6.46:

$$\forall s \in S_{rv}^0(\Delta), \ \forall \sigma \in \Sigma(s),$$

$$y^*(s,\sigma) \leq \{ \min_{(s,\sigma):\, s\in S^0_{rv}(\Delta),\, \sigma\in\Sigma(s)} \tau(s,\sigma)\}^{-1} I_{\{s'\in S_r(\Phi(S);\Delta)\}} \qquad (6.50)$$

The pricing logic underlying Equation 6.50, that facilitates the performance evaluation of the DIS-CON-RAS, $\Phi(S)$, resulting from a single edge selection S, can be subsequently extended to a *Mixed Integer Programming (MIP)* (Winston, 1995) formulation for the entire thinning problem, as follows: Consider an enumeration $\{S_1,\ldots,S_{N'}\}$ of all the '$\preceq$'-*maximal* edge selections satisfying Constraint 6.49, and associate a *binary* variable I_u with each selection S_u, such that $I_u = 1$ indicates that the finally selected edge sets $\mathcal{E}^S_j$ are those in selection S_u. Obviously,

$$\sum_u I_u = 1 \qquad (6.51)$$

In addition, in the spirit of Equation 6.50, we must have:

$$y^*(s,\sigma) \leq \{ \min_{(s,\sigma):\, s\in S^0_{rv}(\Delta),\, \sigma\in\Sigma(s)} \tau(s,\sigma)\}^{-1} \sum_{u:s'\in S_r(\Phi(S_u);\Delta)} I_u, \qquad (6.52)$$
$$\forall s \in S^0_{rv}(\Delta),\ \forall\sigma\in\Sigma(s)$$

i.e., the transition corresponding to control action $\sigma \in \Sigma(s)$ can be activated in the final solution, only if this solution engages an edge selection S_u such that the resulting state s' is in $S_r(\Phi(S_u);\Delta)$. The complete formulation of the thinning problem is obtained by combining the above two constraints with the formulation of Equations 6.45, 6.40–6.42 and 6.46:

$$\max \sum_{s\in S^0_{rv}(\Delta)} \sum_{\sigma\in\Sigma(s)} G_1(s,\sigma)y^*(s,\sigma)$$

s.t.

$$\sum_{\sigma\in\Sigma(s)} y^*(s,\sigma) - \sum_{s'\in S^0_{rv}(\Delta)} \sum_{\sigma\in\Sigma(s')} p(s;s',\sigma)y^*(s',\sigma) = 0,\ \ \forall s \in S^0_{rv}(\Delta);$$

$$\sum_{s\in S^0_{rv}(\Delta)} \sum_{\sigma\in\Sigma(s)} \tau(s,\sigma)y^*(s,\sigma) = 1;$$

$$\sum_{s\in S^0_{rv}(\Delta)} \sum_{\sigma\in\Sigma(s)} G_j(s,\sigma)y^*(s,\sigma) - \rho_j \sum_{s\in S^0_{rv}(\Delta)} \sum_{\sigma\in\Sigma(s)} G_1(s,\sigma)y^*(s,\sigma) = 0,$$
$$\forall j \geq 2;$$

$$y^*(s,\sigma) \leq \{ \min_{(s,\sigma):\, s\in S^0_{rv}(\Delta),\, \sigma\in\Sigma(s)} \tau(s,\sigma)\}^{-1} \sum_{u:s'\in S_r(\Phi(S_u);\Delta)} I_u,$$
$$\forall s \in S^0_{rv}(\Delta),\ \forall\sigma\in\Sigma(s);$$

$$\sum_u I_u = 1;$$

$$y^*(s,\sigma) \geq 0,\ \ \forall s \in S^0_{rv}(\Delta),\ \forall\sigma\in\Sigma(s);$$
$$I_u \in \{0,1\},\ \forall u.$$

This MIP formulation can be solved, in principle, through available commercial solvers (Winston, 1995). However, we notice that the size of both variable sets, $y^*(s, \sigma)$ and I_u, is a super-polynomial function of $|\Phi|$ and, in general, these sets will grow very fast. Therefore, for practical purposes, it is important to develop alternative approximating schemes to it. Such approximating schemes can be based on the adaptation, for the considered problem, of typical search-based combinatorial optimization algorithms, like *Tabu search* (Glover, 1990) and *genetic algorithms* (Buckles and Petry (eds.), 1992).

3. Developing computationally efficient approximations of the optimal scheduling policy Ψ^*

The methodology of Section 1.2 provides a systematic and rigorous characterization of the performance-oriented control problem that arises in the context of logically controlled RAS, and of its optimal solution, but, from a computationally standpoint, it is severely limited by the fact that the derived formulations require the exhaustive enumeration of the underlying state space.[10] This is not surprising, since, as it was observed in the opening discussion of this chapter, the scheduling problems arising in the operational context of the considered RAS class are already known to be NP-Hard (Pinedo, 2002; Garey and Johnson, 1979). Thus, similar to the case of the RAS logical control problem, it is imperative for the effective deployment of the RAS control framework of Figure 1.7, that we develop computationally effective and efficient approximations to the optimal scheduling policy Ψ^*. Of course, as it was observed in the opening discussion of this chapter, the already available and extensively used *dispatching rules* can provide a first baseline towards the sought approximations. In this section, we briefly outline some additional research ideas that could capitalize on the optimality characterizations provided in the earlier parts of this chapter, in order to develop more efficient and systematic approximations of the optimal policy Ψ^*. Another dimension of the proposed research plan seeks the enhancement of the prevailing, dispatching rule-based practice, through the provision of a set of computational tools that can potentially identify, for a given RAS configuration, the dispatching rules that hold the highest promise. All the ideas discussed below, are based on recently developed results in the relevant fields of *approximate dynamic programming* (Bertsekas and Tsitsiklis, 1996) and *queueing network theory* (Chen and Yao, 2001); while the applicability and value of these ideas for the considered problem remains to be proven, their systematic investigation holds also considerable potential for the aforementioned disciplines, since it provides a challenging testing field for

---

[10]In fact, things are even worse, since the aforementioned formulations require also the enumeration of the entire set of state-action pairs!

them, and also opportunity for expansion through specialization of their more general results to the considered application context.

Developing computationally efficient approximations of Ψ^* through neurodynamic programming.   As it was seen in the discussion of the CTMDP problem considered in Section 1, a solution approach to it seeks to compute first the optimal relative value function $h^*(\cdot)$ and the optimal expected average gain V^*, satisfying Bellman's equation 6.31, and subsequently it derives an optimal policy, Ψ^*, from them, through Equation 6.32. In fact, an equivalent but discretized, with respect to time, version of the problem, could have engaged only the corresponding optimal relative value function in the specification of the optimal policy Ψ^*. Similar solution approaches are applicable for most other versions of the MDP problem.   Motivated by these results, *Neuro-dynamic programming (NDP)* (Bertsekas and Tsitsiklis, 1996) is a recently developed field that seeks the approximate solution of MDP problems with very large state spaces, through the approximation of the corresponding *optimal (relative) value function*, $h^*(\cdot)$, by a more compact representation, $\tilde{h}(\cdot; r)$. This representation is known as the approximation *architecture* in the relevant literature, and it essentially constitutes a function that, given any state s, it returns the value $\tilde{h}(s; r)$ as an approximation of $h^*(s)$. The vector r, appearing in $\tilde{h}(\cdot; r)$, is known as the *"weight"* vector of the approximation architecture, and it is priced in a way that minimizes – or more generally, tends to minimize – some distance metric of $\tilde{h}(\cdot; r)$ from its target function $h^*(\cdot)$. The pricing of the weight vector, r, according to the aforementioned logic, is known as the *tuning* or the *fitting* of the architecture $\tilde{h}(\cdot; r)$ to the target function $h^*(\cdot)$; the resulting value of the employed distance metric characterizes the *"goodness-of-fit"* of the approximation. The availability of an approximation architecture, $\tilde{h}(\cdot; r)$, tuned to a particular optimal value function, $h^*(\cdot)$, subsequently allows the specification of an *"on-line"* action selection policy, $\tilde{\Psi}$, through a logic similar to that of Equation 6.32, where $h^*(\cdot)$ is replaced by $\tilde{h}(\cdot; r)$.[11]

The above general description of the neuro-dynamic programming approach indicates that its successful implementation on the performance-oriented control problem of logically controlled RAS will necessitate the development of

1. a pertinent approximation architecture $\tilde{h}(\cdot; r)$, able to generate good approximations of the optimal relative value function $h^*(\cdot)$, and eventually, a policy $\tilde{\Psi}$ leading to a throughput $\tilde{TH}$ that is close to the optimal throughput TH^*;

---

[11] As it was already mentioned, V^* can be eliminated from the policy-defining equation by discretizing the considered MDP problem, with respect to time; c.f. (Puterman, 1994; Bertsekas, 1995b) for the relevant details.

2 a computationally effective and efficient methodology for tuning the parameter vector r, so that the selected approximation architecture $\tilde{h}(\cdot; r)$ provides the best possible fit of the target relative value function $h^*(\cdot)$, for any given RAS configuration.

Regarding the problem of selecting a pertinent approximation architecture, $\tilde{h}(\cdot; r)$, for the RAS performance control problem, we make the following observations: Since the function $\tilde{h}(\cdot; r)$ will be engaged in the real-time computation underlying the specification of policy $\tilde{\Psi}$, the complexity of its structure, including the dimensionality of the weight vector r, must be kept fairly low. More specifically, in order to stay within the design specifications set in Section 3 of Chapter 1, the evaluation of $\tilde{h}(\cdot; r)$ on any given state s must be a task of polynomial complexity with respect to the RAS size $|\Phi|$. On the other hand, constraining the complexity of function $\tilde{h}(\cdot; r)$, especially, its *"degrees of freedom"* provided by the weight vector r, can have an adversarial impact on the representational capability of the architecture, and the eventual quality of the generated approximations. Some critical help in resolving this trade-off can be provided by any analytically and/or empirically derived information regarding the structure of the target function $h^*(\cdot)$, and its involved non-linearities. It is expected that, for the problem under consideration, the burgeoning theory of multi-class queueing networks will be able to provide some relevant useful insights. Indeed, a preliminary study, reported in (Choi, 2004; Choi and Reveliotis, 2004), indicates that, for the case of *single*-process-type RAS,[12] the architecture defined by the weighted sum of some functions[13] of the RAS state, s, that characterize the workload distribution across the system workstations, the available buffering capacity in these stations, and the two or three-order interactions of these quantities, is able to provide a policy $\tilde{\Psi}$ with a performance level that is consistently comparable to the best performance attained through the application of the typically used dispatching rules for these systems. Yet, despite the encouraging nature of this result, there is a host of additional issues that must be addressed with respect to the selection of a good architecture, $\tilde{h}(\cdot; r)$, for the considered problem. For one thing, the aforementioned study was conducted on rather small RAS configurations, that were amenable to the computation of the optimal relative value function, $h^*(\cdot)$, and the resulting optimal throughput, TH^*, through the standard MDP methodologies. Hence, there is a need for validation of the aforementioned results through testing on a broader set of RAS configurations; in particular, it is important to investigate how the performance of the aforementioned architecture scales to larger-sized RAS, where the difference between the dimensionality of the vector function

---

[12]This restriction has allowed us to deal with the *unconstrained* version of the considered CTMDP problem; c.f. the relevant discussion of Section 1.2.

[13]These functions are known as *"feature"* functions, in the NDP terminology.

$h^*(\cdot)$ and the degrees of freedom of the approximating architecture, provided by the weight vector r, becomes very large. Another interesting issue is the systematic assessment of the impact of the distance metric that is employed for measuring the goodness-of-fit of the approximation, $\tilde{h}(\cdot; r)$, of $h^*(\cdot)$, on the performance of the resulting policy. A third issue is the systematic analysis of the significance of the various "feature" functions, employed in the aforementioned architecture, for the quality of the performance of the resulting policy, and the potential identification of additional, more relevant "feature" functions, leading to further improved policies. Finally, a last issue is the assessment of more general approximation architectures, like neural nets, splines, etc., as potential approximation architectures for the considered problem.

Regarding the issue of the weight-tuning methodology that will fit the approximating architecture, $\tilde{h}(\cdot; r)$, to the target relative value function, $h^*(\cdot)$, we notice that it depends, in general, on the structure of (i) the problem under consideration, and (i) the employed architecture $\tilde{h}(\cdot; r)$. Most of the currently available algorithms with theoretically guaranteed correct behavior concern total expected discounted gain MDP's, and they are limited to approximation architectures with a linear dependence on the weight vector r. From a methodological standpoint, they combine dynamic programming, estimation theory and simulation. Extension of these results to expected average gain MDP's and/or more complex approximation architectures seems to be a quite challenging problem. In the light of this remark, the selection of a pertinent architecture that is able to provide good approximations while maintaining a simple analytical structure, is becoming even more important. More generally, it seems that the successful development of an effective weight-tuning methodology for the considered approximation problem will be contingent upon the effective exploitation of any special structure available in the RAS performance control problem under consideration, and the eventually adopted approximation architecture $\tilde{h}(\cdot; r)$.

A last issue that challenges the effective deployment of the aforementioned neuro-dynamic programming methodology on the problem of the performance-oriented control of logically controlled RAS, is the *constrained* nature of the resulting MDP problem, in case of RAS with more than one process types. While the throughput-ratio constraints of Equation 6.43 can be easily added in the LP formulation of Equations 6.26-6.29, it is not clear at all how they can be expressed in the DP algorithms that derive the optimal policy Ψ^* through the computation of the optimal relative value function. Yet, the entire idea of neuro-dynamic programming is based on the specification of the employed scheduling policy, $\tilde{\Psi}$, through the approximation of the optimal relative value function by $\tilde{h}(\cdot; r)$. A potential solution to this complication might be the problem conversion to an unconstrained CTMDP problem, through the redefinition of the problem gains, $G_j(s, \sigma)$, so that they account for the throughput-ratio constraints.

Performance bounds for logically controlled RAS. The works of (Kumar and Kumar, 1994; Bertsimas et al., 1994) have shown that it is possible to derive effective performance bounds of multi-class queueing networks (MCQN) by solving some LP – or, more generally, MP – formulations that are polynomially sized with respect to the number of the net classes. Generally speaking, the constraints appearing in these LP formulations essentially express the existence of a limiting distribution for the stochastic process characterizing the operation of a stable MCQN, and various other restrictions imposed by the net structure and the applied scheduling policy. More recently, the works of (Reveliotis and Ferreira, 1997; Jeng et al., 2000) have established that the aforementioned methodology can be adapted in order to provide performance bounds for the class of SU-RAS, considered in this work. From a practical standpoint, the availability of good-quality bounds characterizing the expected performance of the considered RAS under various scheduling policies, can allow the (partial) ordering of these policies in terms of their expected performance, and therefore, the identification of good candidate policies for any given RAS configuration. Major research issues along this line are (i) the extension of the methodology so that it encompasses more general operational regimes and scheduling policies, and (ii) the strengthening of the corresponding bounds through the identification of more pertinent constraints for the aforementioned MP formulations.

Characterizing the asymptotic performance of popular scheduling policies. A last issue concerns the characterization of the performance of various scheduling policies / dispatching rules, in the limiting case that the capacities of the RAS resources take very large, but still finite values. This question is relevant since past work on multi-class queueing networks with infinite buffering capacity at all stations, has managed to establish the optimality of certain, quite popular dispatching rules (Kumar, 1994a); it would be interesting to see how these results relate to the performance obtained by the application of these policies on the aforementioned RAS. From a methodological standpoint, the limiting regimes of *fluid*, and possibly, *Brownian motion*-based models (Dai, 1998; Chen and Yao, 2001), seem to be the most effective tools for carrying out this analysis.

4. Historical and bibliographical notes

As it was mentioned in earlier parts of this chapter, the integration of logical and performance-oriented control, in the operational context of the RAS classes considered in this book, has received only limited attention from the research community. This is also reflected in the number and the dates of publications available on this topic. Characteristically, the material of Sections 1.2 and 2 is coming from (Reveliotis and Choi, 2004), a work presented at WODES'04. Some additional past works dealing with the characterization of the optimal scheduling policy for logically controlled RAS, and with the impact of the applied logical control policy / strategy on the system performance, can be found in (Choi, 2004; Choi and Reveliotis, 2003; Reveliotis, 2000a; Reveliotis and Choi, 2003). More specifically, the works of (Choi, 2004; Choi and Reveliotis, 2003) seek to provide a systematic characterization of the interaction of the logical and performance control, and of the structure of the optimal scheduling policy, for a single-process-type RAS, motivated by the basic product flow in contemporary semiconductor manufacturing plants. On the other hand, the works of (Reveliotis, 2000a; Reveliotis and Choi, 2003) employ a modelling framework similar to that introduced in Section 1.2, in order to study the systematic selection of a deadlock resolution strategy for Markovian DIS-CON-RAS that optimizes some typical performance index like the RAS throughput. Hence, these two works rationalize, for the addressed operational context, the selection between the two major approaches of deadlock avoidance and deadlock detection & recovery, an issue that has been posed as a dilemma in the earlier literature. Finally, we notice that two other works that have sought to integrate the problem of establishing RAS non-blocking behavior with the corresponding performance control / scheduling problem, are those presented in (Ramaswamy and Joshi, 1996; Mati et al., 2001). Both of these works address the aforementioned problem in an SU-RAS context modelling the operation of contemporary flexibly automated production environments; assuming deterministic stage processing times and a priori knowledge of the entire production requirements over a certain planning horizon, they proceed to develop off-line an optimized production schedule that is also deadlock-free.

Next we briefly discuss each of the major modelling frameworks that was employed, or, more generally, addressed, during the chapter development, highlighting the main trends in them, and providing some introductory references. The theory of stochastic processes, in general, and of Markov processes, in particular, are very well-established fields in mathematical literature. Our coverage of this topic was based on the introductory material of (Cassandras and Lafortune, 1999), but there is a vast literature in the area; some indicative references are those of (Ross, 1983; Gallager, 1996). Markov Decision Processes and the associated field of Dynamic Programming is another well-established and quite active research discipline; excellent expositions of the major results in this area can be found in (Puterman, 1994; Bertsekas, 1995a; Bertsekas, 1995b). On the other hand, as it was mentioned in Section 3, Neuro-dynamic programming is a rather novel approach to approximate dynamic programming. In fact, the origins of the method can be traced in an effort initiated in Artificial Intelligence, in the early 80's, known as *Reinforcement Learning*. Reinforcement learning sought to develop the mathematical apparatus that would enable an intelligent agent to improve its performance on a certain task, by exploiting some evaluative feedback, communicated to the agent by its operational environment, in an "on-line" mode. However, it was the association of this initiative with the MDP framework and Dynamic programming, that provided to the area a rigorous mathematical basis, and brought the entire discipline closer to mathematical systems theory and engineering. An excellent treatment of reinforcement learning and the AI perspective of neuro-dynamic programming is provided in (Sutton and Barto, 1998), while a more formal treatment of the mathematical underpinnings of neuro-dynamic programming can be found in (Bertsekas and Tsitsiklis, 1996). (Multi-class) queueing network theory has emerged as a separate area of the fields of applied probability and stochastic processes, in an effort to develop the necessary analytical and computational tools for the performance modelling and control of a number of technological applications, in which a number of concurrently executing processes compete for the acquisition of some limited processing capacity. In particular, multi-class queueing network theory has received considerable interest through the advent of contemporary manufacturing, telecommunication, and data-processing networks. The material of (Yao (Ed.), 1994) provides a quite encompassing exposition of the active research topics,[14] while (Chen and Yao, 2001) offers a textbook-like treatment of some major recent developments in the area.

---

[14]However, for many of these topics, the indicated state-of-art is quite outdated.

Chapter 7

EPILOGUE

The primary thesis underlying the development of this book has been the observation that many contemporary technological applications have failed to take full advantage of their inherent flexibility, due to the lack of a control paradigm that will manage effectively the resultant system behavior with respect to the allocation of its resources. Hence, the book has set out to provide this missing control paradigm, by building upon recently emerged results in modern control theory, especially an area of it known as Discrete Event Systems (DES). The fundamental ideas characterizing the proposed control paradigm are epitomized by the control architecture depicted in Figure 1.7. A discerning feature of this architecture has been its emphasis on (i) the robustness of the control function to the system stochasticities and the various operational contingencies, (ii) its scalability so that it applies to the large-scale context of the target technological applications, and, of course, (iii) the operational efficiency of the controlled system. In its effort to support these attributes, the proposed architecture is based on a closed-loop approach to the underlying control problem, but also, to a systematic decomposition of the control function, to one component seeking the logical correctness and consistency of the system behavior, and another component addressing performance considerations; the first of these components was characterized as the logical controller of the underlying resource allocation function, while the second component was characterized as the performance-oriented controller. The main part of the book was devoted to the rigorous characterization of the control problems addressed by each of these two controllers, and also, the interaction of these two problems. A notion of optimal control was formulated for both cases, and it was shown that the corresponding optimal control policy has a computational complexity that would not permit its effective real-time implementation on most "real-world" application contexts. Thus, another major contribution of this book has been the development of effective and computationally efficient approximations of the optimal control policy, especially for the novel problem of RAS logical control.

The ultimate objective is that the aforementioned control architecture will function as a formal basis for the development of the "next-generation operating systems", which will be able to support the robust, yet highly flexible, and efficient operation of the target technological applications; in particular, we believe that modern flexibly automated (e..g, semiconductor) manufacturing, autonomous transportation systems, and also, modern workflow management systems can highly benefit from such an undertaking. We also believe that this undertaking will be highly facilitated by an orchestrated effort aiming at the development of a set of computational "toolboxes" that will enable the expedient deployment of the aforementioned control function on any given application context. While this effort can be initiated and led by the relevant research community, interaction with, and understanding of, the target industries is of paramount importance for the eventual acceptance of the final product.

Finally, it must also be noticed that additional work is needed towards complementing and strengthening the analytical results presented in this book. For instance, with respect to the RAS logical control problem, it would be interesting to identify additional "special" RAS structure that admits optimal non-blocking supervision of polynomial complexity, especially for RAS classes other than SU-RAS. Providing an analytical characterization of the conditions that must be satisfied by a matrix A in order to be able to provide correct algebraic supervision with respect to any given RAS configuration, is another interesting theoretical issue, with significant practical implications for the design of such policies.[1] Even more importantly, the research program outlined in Section 3, regarding the systematic development of approximations to the optimal scheduling policy, Ψ^*, of logically controlled RAS, that will be more efficient than the currently employed dispatching rules, is a challenging missing link to the completion of the relevant theory. Effective and efficient contingency accommodation is another topic that currently has received very limited attention in the relevant literature; as it was discussed in Chapter 4, the currently available results, especially with respect to proactive approaches, are of a very exploratory nature, and they apply to a very limited set of cases. In addition, there are no results connecting the problem of contingency accommodation to the RAS performance. Finally, another topic of, both, theoretical and practical interest, is the distribution of the control function in a way that it reflects the physical organization of the underlying application, especially for very large-

---

[1] Some very interesting preliminary results on this problem were developed while this book was ready to go to the press, and they can be traced at S. Reveliotis and L. Tian, *"Implicit Siphon Control and its Implications for the Liveness Enforcing Supervision of Sequential Resource Allocation Systems"*, Tech. Report, School of Industrial & Systems Eng., Georgia Tech, Fall 2004; these results will also be submitted to the 26th International Conference on Application and Theory of Petri nets.

scale application contexts. Space limitations did not allow us to touch upon this problem in the book development, but a prototypical characterization of it, in the context of contemporary flexibly automated production environments, can be found in (Park et al., 2002).

GLOSSARY

AGV   Automated Guided Vehicle

AI   Artificial Intelligence

AMG   Augmented Marked Graph

BK-DAA   Banaszak & Krogh's Deadlock Avoidance Algorithm

cdf   cumulative distribution function

c.f.   cross-reference

CON   Conjunctive

COR   Coordinating

CPX   Complex

CTMC   Continuous-Time Markov Chain

CTMDP   Continuous-Time Markov Decision Process

DAP   Deadlock Avoidance Policy

DES   Discrete Event System

DIS   Disjunctive

DP   Dynamic Programming

DTMC   Discrete-Time Markov Chain

E-DIS-CON-RAS   Extended DIS-CON-RAS

FMS   Flexible Manufacturing System

FSA   Finite State Automaton

G-RUN   Generalized Resource Upstream Neighborhood (policy)

GSMP   Generalized Semi-Markov Process

I/O   Input / Output

IE   Industrial Engineering

IEEE   Institute of Electrical and Electronics Engineers

IIE   Institute of Industrial Engineers

i.e.   id est

iff   if and only if

IP   Integer Program *or* Integer Programming

IT   Information Technology

LES   Liveness Enforcing Supervision *or* Liveness Enforcing Supervisor

LIN   Linear

LP   Linear Program *or* Linear Programming

MC   Markov Chain

MDP   Markov Decision Process

MCQN   Multi-Class Queueing Network

MIP   Mixed Integer Program *or* Mixed Integer Programming
MP   Mathematical Program *or* Mathematical Programming
NDP   Neuro-Dynamic Programming
OR   Operations Research
PK   Polynomial Kernel
PN   Petri Net
QN   Queueing Network
R&W   Ramadge & Wonham
RAS   Resource Allocation System
RDG   Resource Dependency Graph
resp. respectively
RL   Reinforcement Learning
RO   Resource Ordering (policy)
RUN   Resource Upstream Neighborhood (policy)
SAT   Satisfiability (problem)
SC   Supervisory Control
SCP   Supervisory Control Policy
ST   Single-Type
s.t.   such that *or* subject to *(depending on the context)*
STD   State Transition Diagram
SU   Single-Unit
U-G-RUN   A G-RUN version for RAS with Type-1 uncontrollability
WFMS   Workflow Management System
WIP   Work-In-Process
WPD   Working Procedure Digraph
w.r.t. with respect to

References

Ahuja, R. K., Magnanti, T. L., and Orlin, J. B. (1993). *Network Flows: Theory, Algorithms and Applications*. Prentice Hall, Englewood Cliffs, NJ.

Araki, T., Sugiyama, Y., and Kasami, T. (1977). Complexity of the deadlock avoidance problem. In *2nd IBM Symp. on Mathematical Foundations of Computer Science*, pages 229–257. IBM.

Askin, R. G. and Goldberg, J. B. (2002). *Design and Analysis of Lean Production Systems*. John Wiley & Sons, Inc., NY, NY.

Banaszak, Z. A. and Krogh, B. H. (1990). Deadlock avoidance in flexible manufacturing systems with concurrently competing process flows. *IEEE Trans. on Robotics and Automation*, 6:724–734.

Banaszak, Z. A. and Roszkowska, E. (1988). Deadlock avoidance in pipeline concurrent processes. *Podstawy Sterowania (Polish Academy of Sciences)*, 18:3–17.

Barkaoui, K. and Ben Abdallah, I. (1995). A deadlock prevention method for a class of fms. In *Proc. of the IEEE Intl Conf. on Systems, Man and Cybernetics*, pages 4119–4124. IEEE.

Barkaoui, K. and Ben Abdallah, I. (1996). Analysis of a resource allocation problem in fms using structure theory of petri nets. In *Proc. of the 1st Intl Workshop on Manufacturing and Petri Nets*, pages 1–15.

Barkaoui, K., Chaoui, A., and Zouari, B. (1997). Supervisory control of discrete event systems based on structure theory of petri nets. In *Proc. of the IEEE Intl Conf. on Systems, Man and Cybernetics*, pages 3750–3755. IEEE.

Bertsekas, D. P. (1995a). *Dynamic Programming and Optimal Control, Vol. 1*. Athena Scientific, Belmont, MA.

Bertsekas, D. P. (1995b). *Dynamic Programming and Optimal Control, Vol. 2*. Athena Scientific, Belmont, MA.

Bertsekas, D. P. and Tsitsiklis, J. N. (1996). *Neuro-Dynamic Programming*. Athena Scientific, Belmont, MA.

Bertsimas, D., Paschalidis, I. Ch., and Tsitsiklis, J. N. (1994). Optimization of multiclass queueing networks: Polyhedral and nonlinear characterizations of achievable performance. *Annals of Applied Probability*, 4:43–75.

Blackstone, J. H., Philips, D. T., and Hogg, G. L. (1982). A state-of-the-art survey of dispatching rules for manufacturing job shop operations. *Int. J. Prod. Res.*, 20:27–45.

Brandin, B. and Wonham, M. W. (1994). Supervisory control of timed discrete event systems. *IEEE Trans. on Automatic Control*, 39:329–342.

Buckles, B. P. and Petry (eds.), F. E. (1992). *Genetic Algorithms*. IEEE Computer Society Press, Los Alamitos, CA.

Cassandras, C. G. and Lafortune, S. (1999). *Introduction to Discrete Event Systems*. Klumwer Academic Pub., Boston, MA.

Chen, H. and Yao, D. D. (2001). *Fundamentals of Queueing Networks*. Springer, NY, NY.

Chew, S. F., Lawley, M., and Reveliotis, S. (2003). Liveness enforcing supervision for resource allocation with complex workflows. In *Proceedings of MMAR'03*, pages 823–829. IEEE.

Choi, J. Y. (2004). *Performance Modelling, Analysis and Control of Capacitated Re-entrant Lines*. PhD thesis, Georgia Institute of Technology, Atlanta, GA.

Choi, J. Y. and Reveliotis, S. A. (2003). A generalized stochastic petri net model for performance analysis and control of capacitated re-entrant lines. *IEEE Trans. on Robotics and Automation*, 19:474–480.

Choi, J. Y. and Reveliotis, S. A. (2004). Relative value function approximation for the capacitated re-entrant line scheduling problems: An experimental investigation. In *Proceedings of CDC'04*, pages –. IEEE.

Chu, F. and Xie, X-L. (1997). Deadlock analysis of petri nets using siphons and mathematical programming. *IEEE Trans. on R&A*, 13:793–804.

Chung, S. L., Lafortune, S., and Lin, F. (1992). Limited lookahead policies in supervisory control of discrete event systems. *IEEE Trans. on Automatic Control*, 37:1921–1935.

Coffman, E. G., Elphick, M. J., and Shoshani, A. (1971). System deadlocks. *Computing Surveys*, 3:67–78.

Cox, D. R. (1955). A use of complex probabilities in theory of stochastic processes. *Proc. of Cambridge Philosophical Society*, 51:313–319.

Dai, J. G. (1998). Stablity of fluid and stochastic processing networks. Technical Report Miscellanea, No. 9, Dept. of Mathematical Sciences, Unversity of Aarhus, Denmark.

Dallery, Y. and Liberopoulos, G. (2000). Extended kanban control system: Combining kabnan and base stock. *IIE Trans.*, 32:369–386.

Desel, J. and Esparza, J. (1995). *Free Choice Petri Nets*. Cambridge Univerrsity Press.

Desrochers, A. A. (1990). *Modeling and Control of Automated Manufacturing Systems*. IEEE Computer Society Press, Washington, DC.

Di Mascolo, M., Frein, Y., and Dallery, Y. (1996). An analytical method for performance evaluation of kanban controlled production systems. *Operations Research*, 44:50–64.

Dijkstra, E. W. (1965). Cooperating sequential processes. Technical report, Technological University, Eindhoven, Netherlands.

Ezpeleta, J., Colom, J. M., and Martinez, J. (1995). A petri net based deadlock prevention policy for flexible manufacturing systems. *IEEE Trans. on R&A*, 11:173–184.

Ezpeleta, J., Tricas, F., Garcia-Valles, F., and Colom, J. M. (2002). A banker's solution for deadlock avoidance in fms with flexible routing and multi-resource states. *IEEE Trans. on R&A*, 18:621–625.

Fanti, M. P., Maione, B., Mascolo, S., and Turchiano, B. (1997). Event-based feedback control for deadlock avoidance in flexible production systems. *IEEE Trans. on Robotics and Automation*, 13:347–363.

Fanti, M. P., Maione, B., and Turchiano, B. (1998a). Deadlock avoidance in cellular manufacturing systems. In *Proceedings of the 1998 IEEE Conference on Systems, Man and Cybernetics*, pages 588–593. IEEE.

Fanti, M. P., Maione, B., and Turchiano, B. (1998b). Event control for deadlock avoidance in production systems with multiple capacity resources. *Studies in Informatics and Control*, 7:343–364.

Fanti, M. P., Maione, B., and Turchiano, T. (2000). Comparing digraph and petri net approaches to deadlock avoidance in fms modeling and performance analysis. *IEEE Trans. on Systems, Man and Cybernetics, Part B*, 30:783–798.

Gaarder, E. H. (1993). Deadlock avoidance in flexible manufacturing systems. Master's thesis, University of Illinois at Urbana-Champaign, Urbana, IL.

Gallager, R. G. (1996). *Discrete Stochastic Processes*. KAP, Boston,MA.

Ganesharajah, T., Hall, N. G., and Sriskandarajah, C. (1998). Design and operational issues in agv-served manufacturing systems. *Annals of OR*, 76:109–154.

Garey, M. R. and Johnson, D. S. (1979). *Computers and Intractability: A Guide to the Theory of NP-Completeness*. W. H. Freeman and Co., New York, NY.

Gates, B. (1995). *The Road Ahead*. Viking, New York, NY.

Gershwin, S. B. (1989). Hierarchical flow control: A framework for scheduling and planning discrete events in manufacturing systems. *Proceedings of the IEEE*, 77:195–209.

Gershwin, S. B. (1994). *Manufacturing Systems Engineering*. PTR Prentice Hall, Englewood Cliffs, N.J.

Ghaffari, A., Rezg, N., and Xie, X. (2003). Design of a live and maximally permissive petri net controller using the theory of regions. *IEEE Trans. on Robotics & Automation*, 19:137–141.

Glasserman, P. and Yao, D. (1994). *Monotone Structure in Discrete-Event Systems*. John Wiley & Sons, Inc., NY,NY.

Glover, F. (1990). Tabu search: A tutorial. *INTERFACES*, 20:74–94.

Gohari, P. and Wonham, M. W. (2000). On the complexity of the supervisory control design in the rw framework. *IEEE Trans. on SMC – Part B*, 30:643–652.

Gold, E. M. (1978). Deadlock prediction: Easy and difficult cases. *SIAM Journal of Computing*, 7:320–336.

Groover, M. P. (1996). *Fundamentals of Modern Manufacturing: Materials, Processes and Systems*. Prentice Hall, Englewood Cliffs, N.J.

Habermann, A. N. (1969). Prevention of system deadlocks. *Comm. ACM*, 12:373–377.

Havender, J. W. (1968). Avoiding deadlock in multi-tasking systems. *IBM Systems Journal*, 2:74–84.

Hillier, F. S. and Lieberman, G. J. (2002). *Introduction to Operations Research (7th ed.)*. McGraw Hill.

Holt, R. D. (1972). Some deadlock properties of computer systems. *ACM Computing Surveys*, 4:179–196.

Hopcroft, J. E. and Ullman, J. D. (1979). *Introduction to Automata Theory, Languages and Computation*. Addison-Wesley, Reading, MA.

Hopp, W. J. and Spearman, M. L. (1996). *Factory Physics: Foundations of Manufacturing Management*. IRWIN, Chicago, IL.

Hsieh, F. S. and Chang, S. C. (1994). Dispatching-driven deadlock avoidance controller synthesis for flexible manufacturing systems. *IEEE Trans. on Robotics and Automation*, 10:196–209.

Iordache, M. V. and Antsaklis, P. J. (2003). Design of t-liveness enforcing supervisors in petri nets. *IEEE Trans. on Automatic Control*, 48:1962–1974.

Jeng, M. and Xie, X. (2001). Modeling and analysis of semiconductor manufacturing systems with degraded behaviors using petri nets and siphons. *IEEE Trans. on Robotics and Automation*, 17:576–588.

Jeng, M., Xie, X., and Hung, W.-Y. (2000). Markovian timed petri nets for performance analysis of semiconductor manufacturing systems. *IEEE Trans. on Systems, Man and Cybernetics – Part B*, 30:757–771.

Jeng, M., Xie, X., and Peng, M. Y. (2002). Process nets with resources for manufacturing modeling and their analysis. *IEEE Trans. on Robotics & Automation*, 18:875–889.

Joshi, S. B., Mettala, E. G., Smith, J. S., and Wysk, R. A. (1995). Formal models for control of flexible manufacturing cells: Physical and system models. *IEEE Trans. on Robotics and Automation*, 11:558–570.

Kumar, P. R. (1994a). Scheduling manufacturing systems of re-entrant lines. In Yao, D. D., editor, *Stochastic Modeling and Analysis of Manufacturing Systems*, pages 325–360. Springer-Verlag.

Kumar, P. R. (1994b). Scheduling semiconductor manufacturing plants. *IEEE Control Systems Magazine*, 14–6:33–40.

Kumar, R. and Garg, V. (1995). *Modeling and Control of Logical Discrete Event Systems*. Kluwer Academic, Pub., Boston, MA.

Kumar, S. and Kumar, P. R. (1994). Performance bounds for queueing networks and scheduling policies. *IEEE Trans. Autom. Control*, 39:1600–1611.

Lawley, M., Reveliotis, S., and Ferreira, P. (1997a). Design guidelines for deadlock handling strategies in flexible manufacturing systems. *Intl. Jrnl. of Flexible Manufacturing Systems*, 9:5–29.

Lawley, M., Reveliotis, S., and Ferreira, P. (1997b). Fms structural control and the neighborhood policy, part 1: Correctness and scalability. *IIE Trans.*, 29:877–887.

Lawley, M., Reveliotis, S., and Ferreira, P. (1997c). Fms structural control and the neighborhood policy, part 2: Generalization, optimization and efficiency. *IIE Trans.*, 29:889–899.

Lawley, M., Reveliotis, S., and Ferreira, P. (1998a). The application and evaluation of banker's algorithm for deadlock-free buffer space allocation in flexible manufacturing systems. *Intl. Jrnl. of Flexible Manufacturing Systems*, 10:73–100.

Lawley, M., Reveliotis, S., and Ferreira, P. (1998b). A correct and scalable deadlock avoidance policy for flexible manufacturing systems. *IEEE Trans. on Robotics & Automation*, 14:796–809.

Lawley, M. and Sulistyono, W. (2002). Robust supervisory control policies for manufacturing systems with unreliable resources. *IEEE Trans. on R&A*, 18:346–359.

Lawley, M. A. (1999). Deadlock avoidance for production systems with flexible routing. *IEEE Trans. Robotics & Automation*, 15:497–509.

Lawley, M. A. and Reveliotis, S. A. (2001). Deadlock avoidance for sequential resource allocation systems: hard and easy cases. *Intl. Jrnl of FMS*, 13:385–404.

Lawrence, P. (1997). *Workflow Handbook*. John Wiley & Sons, Inc., NY, NY.

Li, Y. and Wonham, W. M. (1988). Deadlock issues in supervisory control of discrete event systems. In *Proc. Conf. Inf. Sci. Syst.*, pages 57–63.

Lu, S. H. and Kumar, P. R. (1991). Distributed scheduling based on due dates and buffer priorities. *IEEE Trans. on Aut. Control*, 36:1406–1416.

Mackulak, G. T., Fowler, J. W., and Schoming (eds.), A. (2002). *MASM 2002*. Tempe, AZ.

Marinescu, D. C. (2002). *Internet-Based Workflow Management: Towards a Semantic Web*. Wiley Interscience, NY,NY.

Mati, Y., Rezg, N., and Xie, X. (2001). Geometric approach and taboo search for scheduling flexible manufacturing systems. *IEEE Trans. o R&A*, 17:805–818.

Medhi, J. (1991). *Stochastic Models in Queueing Theory*. Academic Press, San Diego, CA.

Moody, J. O. and Antsaklis, P. J. (1998). *Supervisory Control of Discrete Event Systems using Petri nets*. Kluwer Academic Pub., Boston, MA.

Murata, T. (1989). Petri nets: Properties, analysis and applications. *Proceedings of the IEEE*, 77:541–580.

Nahmias, S. (1997). *Production and Operations Analysis – 3rd ed.* IRWIN, Chicago, IL.

Naylor, A. W. and Volz, R. A. (1987). Design of integrated manufacturing system control software. *IEEE Trans. SMC*, 17:881–897.

Panwalkar, S. S. and Iskander, W. (1977). A survey of scheduling rules. *Oper. Research*, 25:45–61.

Papadopoulos, H. T., Heavy, C., and Browne, J. (1993). *Queueing Theory in Manufacturing Systems Analysis and Design*. Chapman & Hall, New York, NY.

Park, J. (2000). *Structural Analysis and Control of Resource Allocation Systems using Petri nets.* PhD thesis, Georgia Institute of Technology, Atlanta, GA.

Park, J. and Reveliotis, S. (2000). Algebraic synthesis of efficient deadlock avoidance policies for sequential resource allocation systems. *IEEE Trans. on R&A*, 16:190–195.

Park, J. and Reveliotis, S. (2001a). Algebraic deadlock avoidance policies for conjunctive / disjunctive resource allocation systems. In *Proc. of ICRA'01.* IEEE.

Park, J. and Reveliotis, S. (2002a). Liveness-enforcing supervision for resource allocation systems with uncontrollable events and forbidden states. *IEEE Trans. on R&A*, 18:234–240.

Park, J. and Reveliotis, S. (2002b). Policy mixtures: A novel approach for enhancing the operational flexibility of resource allocation systems with alternate routings. *IEEE Trans. on R&A*, 18:616–620.

Park, J., Reveliotis, S., Lawley, M., and Ferreira, P. (2001). Correction on the run dap for conjunctive ras presented in "polynomial comlpexity deadlock avoidance policies for sequential resource allocation systems". *IEEE Trans. on Automatic Control*, 46:672.

Park, J. and Reveliotis, S. A. (2001b). Deadlock avoidance in sequential resource allocation systems with multiple resource acquisitions and flexible routings. *IEEE Trans. on Automatic Control*, 46:1572–1583.

Park, J., Reveliotis, S. A., Bodner, D., and McGinnis, L. (2002). A distributed event-driven control architecture for flexibly automated manufacturing systems. *Intl. Jrnl on CIM*, 15:109–126.

Park, S. and Lim, J. (1999). Fault-tolerant robust supervisor for discrete event systems with model uncertainty and its application to a workcell. *IEEE Trans. on R&A*, 15:386–391.

Perkins, J. R. and Kumar, P. R. (1995). Optimal control of pull manufacturing systems. *IEEE Trans. Automat. Control*, 40:2040–2051.

Perros, H. G. (1994). *Queueing Networks with Blocking: Exact and Approximate Solutions.* Oxford University Press, NY, NY.

Peterson, J. L. (1981). *Operating System Concepts.* Addison-Wesley.

Pinedo, M. (2002). *Scheduling.* Prentice Hall, Upper Saddle River, NJ.

Puterman, M. L. (1994). *Markov Decision Processes: Discrete Stochastic Dynamic Programming.* John Wiley & Sons.

Ramadge, P. J. G. and Wonham, W. M. (1989). The control of discrete event systems. *Proceedings of the IEEE*, 77:81–98.

Ramaswamy, S. E. and Joshi, S. B. (1996). Deadlock-free schedules for automated manufacturing workstations. *IEEE Trans. on R&A*, 12:391–400.

Reveliotis, S. A. (1996). *Structural Analysis & Control of Flexible Manufacturing Systems with a Performance Perspective.* PhD thesis, University of Illinois, Urbana, IL.

Reveliotis, S. A. (1999). Accommodating fms operational contingencies through routing flexibility. *IEEE Trans. on R&A*, 15:3–19.

Reveliotis, S. A. (2000a). An analytical investigation of the deadlock avoidance vs. detection & recovery problem in buffer-space allocation of flexibly automated production systems. *IEEE Trans. on SMC: Part B*, 30:–.

Reveliotis, S. A. (2000b). Conflict resolution in agv systems. *IIE Trans.*, 32(7):647–659.

Reveliotis, S. A. (2003a). On the siphon-based characterization of liveness in sequential resource allocation systems. In *Applications and Theory of Perti Nets 2003*, pages 241–255.

Reveliotis, S. A. (2003b). Structural analysis of assembly/disassembly resource allocation systems. In *Proc. of ICRA 2003.* IEEE.

Reveliotis, S. A. and Choi, J. Y. (2003). On the optimality of randomized deadlock avoidance policies. *Jrnl of DEDS: Theory and Applications*, 13:303–320.

Reveliotis, S. A. and Choi, J. Y. (2004). The thinning problem. In *Proceedings of WODES'04*, pages –. IFAC.

Reveliotis, S. A. and Ferreira, P. M. (1996). Deadlock avoidance policies for automated manufacturing cells. *IEEE Trans. on Robotics & Automation*, 12:845–857.

Reveliotis, S. A. and Ferreira, P. M. (1997). A polynomial complexity tool for evaluating the performance of structurally controlled fms. In *1997 IEEE International Conference on Robotics and Automation*, pages –. IEEE Robotics and Automation Society.

Reveliotis, S. A., Lawley, M. A., and Ferreira, P. M. (1997). Polynomial complexity deadlock avoidance policies for sequential resource allocation systems. *IEEE Trans. on Automatic Control*, 42:1344–1357.

Rohloff, K. and Lafortune, S. (2003). Recent results on computational issues in supervisory control. In *ATPN-Workshop on Discrete Event Systems Control*, pages 15–35. INRIA/Univ. of Osaka.

Ross, S. M. (1983). *Stochastic Processes*. Wiley, N.Y.

Roszkowska, E. (2003). Liveness of states in closed agv systems with dynamic routing. In *Proceedings of MMAR'03*, pages –. IEEE.

Roszkowska, E. (2004). Liveness enforcing in closed agv systems with dynamic routing. In *Proceedings of ICRA'04*, pages –. IEEE.

Roszkowska, E. and Wojcik, R (1993). Problems of process flow feasibility in fas. In *CIM in Process and Manufacturing Industry*, pages 115–120. Pergamon Press.

Rust, K., Wright, R., and Shopbell, M. (2002). Comparative analysis of 300mm automated material handling systems (amhs). In *MASM 2002*, pages 240–245.

Silva, M., Teruel, E., and Colom, J. M. (1998). Linear algebraic and linear programming techniques for the analysis of place/transition net systems. In Reisig, W. and Rozenberg, G., editors, *Lecture Notes in Computer Science, Vol. 1491*, pages 309–373. Springer-Verlag.

Singer, P. (1995). The driving forces in cluster tool development. *Semiconductor International*, July '95:113–118.

Suarez, F. F., Cusumano, M. A., and Fine, C. H. (1997). An empirical study of manufacturing flexibility in printed circuit board assembly. *Operations Research*, 44:223–240.

Sulistyono, W. and Lawley, M. (2002). Robust supervisory control for manufacturing systems with unreliable resources. In *Proceeding of ICRA'02*, pages 199–204. IEEE.

Sutton, R. S. and Barto, A. G. (1998). *Reinforcement Learning*. The MIT Press, Cambridge, MA.

Tricas, F., Colom, J. M., and Ezpeleta, J. (1999). A solution to the problem of deadlock in concurrent systems using petri nets and integer linear programming. In *Proceedings of the 11th Eurpoean Simulation Symposium*, pages 542–546.

Tricas, F., Garcia-Valles, F., Colom, J. M., and Ezpeleta, J. (1998). A structural approach to the problem of deadlock prevention in processes with resources. In *Proceedings of the 4th Workshop on Discrete Event Systems*, pages 273–278. IEE.

Van der Aalst, W. (1996). Structural characterizations of sound workflow nets. Technical Report Computing Science Reports 96/23, Eindhoven University of Technology.

Van der Aalst, W. (1997). Verification of workflow nets. In Azema, P. and Balbo, G., editors, *Lecture Notes in Computer Science, Vol. 1248*, pages 407–426. Springer Verlag.

Van der Aalst, W. and Van Hee, K. (2002). *Workflow Management: Models, Methods and Systems*. The MIT Press, Cambridge, MA.

Van Hee, K., Sidorova, N., and Voorhoeve, M. (2003). Soundness and separability of workflow nets in the stepwise refinement approach. In Van der Aalst, W. and Best, E., editors, *Lecture Notes in Computer Science, Vol. 2679*, pages 337–356. Springer Verlag.

Viswanadham, N. and Narahari, Y. (1992). *Performance Modeling of Automated Manufacturing Systems*. Prentice Hall, Englewood Cliffs, NJ.

Viswanadham, N., Narahari, Y., and Johnson, T. L. (1990). Deadlock avoidance in flexible manufacturing systems using petri net models. *IEEE Trans. on Robotics and Automation*, 6:713–722.

Winston, W. L. (1995). *Introduction To Mathematical Programming: Applications and Algorithms, 2nd ed.* Duxbury Press, Belmont, CA.

Wu, N. and Zhou, M. (2001). Avoiding deadlock and reducing starvation and blocking in automated manufacturing systems. *IEEE Trans. on R&A*, 17:658–669.

Wysk, R. A., Yang, N. S., and Joshi, S. (1991). Detection of deadlocks in flexible manufacturing cells. *IEEE Trans. on Robotics and Automation*, 7:853–859.

Wysk, R. A., Yang, N. S., and Joshi, S. (1994). Resolution of deadlocks in flexible manufacturing systems: Avoidance and recovery approaches. *Journal of Manufacturing Systems*, 13:128–138.

Xie, X. and Jeng, M. (1999). Ercn-merged nets and their analysis using siphons. *IEEE Trans. on R&A*, 13:692–703.

Xing, K. Y., Hu, B. S., and Chen, H. X. (1996). Deadlock avoidance policy for petri net modeling of flexible manufacturing systems with shared resources. *IEEE Trans. on Aut. Control*, 41:289–295.

Yalcin, A. and Boucher, T. (2000). Deadlock avoidance in flexible manufacturing systems using finite automata. *IEEE Trans. on Robotics & Automation*, 16:424–428.

Yao (Ed.), D. D. (1994). *Stochastic Modeling and Analysis of Manufacturing Systems.* Springer-Verlag, NY, NY.

Zhou, M. and DiCesare, F. (1993). *Petri Net Synthesis for Discrete Event Control of Manufacturing Systems.* Kluwer Academic Pub., Boston, MA.

Zhou, M. and Fanti, M. P. (2004). *Deadlock Resolution in Computer-Integrated Systems.* Marcel Dekker, Inc., Singapore.

About the Author

Dr. Spyros A. Reveliotis received his Ph.D. in Industrial Engineering from the University of Illinois at Urbana-Champaign in 1996. He also holds a B.Sc. degree in Electrical Engineering, and an M.Sc. degree in Computer Systems Engineering. Since Fall 1996, he has been a faculty member of the School of Industrial & Systems Engineering, at the Georgia Institute of Technology, where he is currently an Associate Professor.

Dr. Reveliotis' research interests lie in the area of Discrete Event Systems theory and its application to the formal modelling, analysis and control of contemporary technological applications. His work has been published in many prestigious journals, including the IEEE Trans. on Automatic Control, IEEE Trans. on Robotics & Automation, IEEE Trans. on Systems, Man & Cybernetics, IIE Trans., the International Journal on Discrete Event Systems, and the International Journal of Flexible Manufacturing Systems. He is a senior member of IEEE, and a member of INFORMS and IIE. He has been an Associate Editor for the IEEE Trans. on Robotics and Automation, and he was also the recipient of the Kayamori Best Paper Award at the 1998 International Conference on Robotics & Automation.

Index

240

242

Early Titles in the
INTERNATIONAL SERIES IN
OPERATIONS RESEARCH & MANAGEMENT SCIENCE
Frederick S. Hillier, Series Editor, *Stanford University*